T0188416

Understanding Mathematical and Statistical Techniques in Hydrology

Understanding Mathematical and Statistical Techniques in Hydrology

An Examples-Based Approach

Harvey J. E. Rodda

Max A. Little

WILEY Blackwell

This edition first published 2015 © 2015 by Harvey J. E. Rodda and Max A. Little

Registered Office
John Wiley & Sons, Ltd, The Atrium, Southern Gate, Chichester, West Sussex, PO19 8SQ, UK

Editorial Offices
9600 Garsington Road, Oxford, OX4 2DQ, UK
The Atrium, Southern Gate, Chichester, West Sussex, PO19 8SQ, UK
111 River Street, Hoboken, NJ 07030-5774, USA

For details of our global editorial offices, for customer services and for information about how to apply for permission to reuse the copyright material in this book please see our website at www.wiley.com/wiley-blackwell.

The right of the author to be identified as the author of this work has been asserted in accordance with the UK Copyright, Designs and Patents Act 1988.

All rights reserved. No part of this publication may be reproduced, stored in a retrieval system, or transmitted, in any form or by any means, electronic, mechanical, photocopying, recording or otherwise, except as permitted by the UK Copyright, Designs and Patents Act 1988, without the prior permission of the publisher.

Designations used by companies to distinguish their products are often claimed as trademarks. All brand names and product names used in this book are trade names, service marks, trademarks or registered trademarks of their respective owners. The publisher is not associated with any product or vendor mentioned in this book.

Limit of Liability/Disclaimer of Warranty: While the publisher and author(s) have used their best efforts in preparing this book, they make no representations or warranties with respect to the accuracy or completeness of the contents of this book and specifically disclaim any implied warranties of merchantability or fitness for a particular purpose. It is sold on the understanding that the publisher is not engaged in rendering professional services and neither the publisher nor the author shall be liable for damages arising herefrom. If professional advice or other expert assistance is required, the services of a competent professional should be sought.

Library of Congress Cataloging-in-Publication Data

Rodda, Harvey.
  Understanding mathematical and statistical techniques in hydrology : an examples-based approach / Harvey Rodda, Max Little.
    1 online resource.
  Includes index.
  Description based on print version record and CIP data provided by publisher; resource not viewed.
  ISBN 978-1-119-07659-9 (pdf) – ISBN 978-1-119-07660-5 (epub) – ISBN 978-1-4443-3549-1 (cloth)
1. Hydrology–Mathematical models.   2. Hydrology–Statistical methods.   I. Title.
  GB656.2.M33
  551.4801′51–dc23

                                  2015023562

A catalogue record for this book is available from the British Library.

Wiley also publishes its books in a variety of electronic formats. Some content that appears in print may not be available in electronic books.

Cover image: The River Avon at Upavon, Wiltshire, UK © Harvey J. E. Rodda

Set in 9.5/13pt Meridien by SPi Global, Pondicherry, India
Printed and bound in Malaysia by Vivar Printing Sdn Bhd

1   2015

Contents

Preface

Understanding Mathematical and Statistical Techniques in Hydrology: An Examples-Based Approach is primarily intended as a textbook to assist undergraduate and postgraduate students with courses or modules in hydrology. In higher education, hydrology as a subject is not usually taught in its entirety as a separate course at undergraduate level but is generally included as a module of geography, environmental science or earth science courses. It can also be included in civil engineering courses which deal with river engineering, drainage, water supply and sewage treatment. More specialized postgraduate courses such as water resources management focus on hydrology. Such undergraduate and postgraduate courses do not generally include any supplementary mathematics and in many cases an advanced school leaving qualification in mathematics is not an essential entry requirement. However, many of the current hydrology textbooks for undergraduate and postgraduate courses assume a high level of mathematical expertise, such as that attained when studying for a mathematics degree. For example, textbooks often present a sequence of differential equations which are impossible to comprehend without having this high level of mathematical knowledge. Instead of assisting the students with their studies these texts when full of mathematical notations are of little interest to the reader. They can also distance students from using mathematics to the extent that they are discouraged from attempting any mathematical-based questions in final exams.

It is commonly the case that students would choose to study hydrology because of their interest in the natural environment, rivers, the hydrological cycle and the human impact. A major part of this is actually going out into the field to observe what is happening, taking measurements and with students often getting their feet wet. Students would not choose to study hydrology because of the chance to sit at a desk solving equations; this would be reserved for students wishing to study mathematics, statistics or a particularly theoretical science. It is never the case that a university would advertise a hydrology course or module as the opportunity to study complex mathematics. However, mathematics is becoming a more integral part of hydrology and other environmental sciences, with the need to explain and quantify many of the basic processes through the use of equations. This is particularly evident in recent years as advances in computing have increased the opportunity for the collection, storage and analysis of data.

Assignments and exam questions from hydrology modules or similar subjects will often have some mathematical components, and university professors particularly from a physical sciences background have a tendency to test students in this subject to demonstrate their understanding of mathematics. An extreme example was when students were set some particularly difficult coursework as part of an engineering geology module. The task was to rearrange a complicated formula to show how a particular parameter was related to other parameters. It was purely an exercise in algebra and did not require any knowledge of engineering geology, just competence at mathematics. The majority of the class felt cheated that they were being assessed on their mathematical ability rather than their understanding of the subject they were studying. For the academic staff member who set the coursework, they may have been emphasizing the fact that not only were they an expert in engineering geology but also highly adept at mathematics, and there are some students who may well excel in a similar way, and end up themselves as academics. However, for most students who are just pleased to leave university with a degree and then look for work in a related field, we hope that assistance with this rather unfair practice of testing them purely on their mathematical ability, rather than the subject which they are studying, can be sought from the content of this book.

Many exams for hydrology courses will have the option for students to attempt one of two questions, where one is a purely written answer and the other always includes some form of calculation. This calculation may well be a simple application of an equation which is provided, so all the students would need to do would be to plug the values into the equation and use their calculator to get an answer. This actual calculation component would only form part of the question, and at least half of the marks would come from a written discussion about the results and the application of the equation. However, it is not uncommon for students to avoid attempting such a question because of their lack of confidence in mathematics. It should not be a requirement that students of hydrology should be able to rearrange any equation which is put in front of them but simply be able to use equations with confidence. The complete avoidance of anything mathematical in exams is largely the result of no adequate texts from which these students could gain the selected hints and tips on the mathematical side. This ability to use mathematics and equations with more confidence would also put them in a better position with prospective employers.

There is perhaps a perception in the scientific community that students who have studied hydrology as part of a geography, environmental science or earth science course are less desirable for employers particularly in the research sector than students who have studied a more theoretical science or maths. The perceived mathematical ability is often given more credibility than the overall understanding of the subject. An example of this was where a candidate for a position in water quality modelling at a UK research institute, who had a PhD in hydrology, specializing in water quality modelling, was rejected in favour of a candidate, who had a PhD in nuclear physics. The selection was made on the basis that the latter would have a better understanding of the mathematical aspects of the research. For employers to take such a position would appear ridiculous. The fact that a nuclear physicist would have no idea of the rudimentary issues in water quality did not seem to be an issue. Such a recruitment policy would rarely be accepted in other fields; would a candidate with a PhD in water quality gain a position at a nuclear physics research establishment if their level of maths was deemed to be higher than competitive candidates with PhDs in nuclear physics? The perception that students who have studied hydrology do not have a particularly strong mathematical ability would originate from the aforementioned problems of a lack of introductory maths within hydrology courses and that students without advanced level maths would perform poorly or look to avoid mathematical-based tasks. With better teaching of mathematics within hydrology courses, candidates for hydrology positions would have more confidence in mathematics and therefore give a better impression to perspective employers.

Outside of education, professional hydrologists are also faced with mathematical challenges often relating to new modelling techniques which have been introduced or new ways of analysing data. Mathematical terminology is rarely questioned and hydrologists will probably have sat through numerous presentations at conferences where mathematical terms have been talked about without any proper definition such as Bayesian, beta distributions and wavelet analysis. Papers in hydrological journals have become more modelling and analysis based and less about measurements and observations. The reason though is quite simple: due to the need to publish, it is cheaper and more efficient to report on some aspect of modelling rather than waiting years for the results of a field experiment. Some papers are now so mathematical, presenting discussions about parameter optimization, uncertainty and measurements of model performance that little is written on the hydrological aspects of study. These are all areas where hydrologists would benefit from a text where mathematical techniques are explained at a level which can be easily understood for those without university or even advanced high school level maths.

How to use this book

A total of six chapters are included in this book, the idea being that they each consider different but none the less related topics which are found within hydrology. Any chapter can therefore be used as reference material without the need for reading any other chapters.

Chapter 1. Fundamentals
The introductory section provides the basic mathematical theory behind the material which is presented in later chapters. This includes a summary of mathematical techniques from the simple use of numbers and operations, the application of algebra and rearranging of equations, the use of functions, a description of calculus and differential equations and a definition of probability and the use of statistics. The content of this chapter does not include specific hydrological examples but is intended to be used as a reference section for the other chapters.

Chapter 2. Statistical modelling
In this chapter, the concept of return period is defined and explained within the broader context of probability and extreme value analysis. It is still the case that often the most common requirement for a practicing hydrologist is to estimate the return period of a particular quantity such as flow or rainfall or to estimate the magnitude of that quantity for a particular return period (e.g. the 1 in 100 year flow). The content includes techniques of statistical modelling, in particular, extreme value analysis such as Gumbel and Weibull model fitting, flood frequency curves and the relationship between return period and flow.

Chapter 3. Mathematics of hydrological processes
This chapter will explain widely used equations of various levels of complexity in physical process hydrology from simple mass-balance equations, the use of exponents, advanced notation and differential equations. It will also explain how these equations can be simplified and rearranged as is often presented in textbooks with a proper written explanation of how the different steps in the rearranging process are undertaken.

Chapter 4. Techniques based on data fitting
This chapter will consider techniques for establishing equations and relationships between observed hydrological variables based on data from field experiments or monitoring such as the work done in the United Kingdom for the *Flood Estimation*

Handbook and similar studies. This will cover line and curve fitting, with differing levels of complexity from simple linear relationships, various non-linear relationships, multiple regression and cyclical patterns.

Chapter 5. Time series data

The problem of time-dependant data is presented in this chapter. Examples of such data are provided over different scales and the analysis of such data is included for identifying trends, smoothing and filtering, and predicting future outcomes. A section on the problems of such analysis where the conditions have been changing over time is included through the question of non-stationarity, and finally the use of modelling based purely on patterns in the data is presented.

Chapter 6. Measures of model performance, uncertainty and stochastic modelling

This chapter includes a particularly important section on how well models can perform, which is often overlooked in the mathematical analysis of hydrological data. A selection of mathematical-based performance measures are presented with examples of performance from hypothetical models. The chapter also includes a definition and discussion of the idea of uncertainty, again an important component in relation to modelling. Finally as a related topic, the development of stochastic modelling is presented which takes many of the ideas of uncertainty analysis to a more practical level.

CHAPTER 1

Fundamentals

1.1 Motivation for this book

Hydrology is the study of water, and in the International Glossary of Hydrology (UNESCO/WMO 1992) it is defined as 'Science that deals with the waters above and below the land surfaces of the Earth, their occurrence, circulation and distribution, both in space and time, their biological, chemical and physical properties, their reaction with their environment, including their relation to living beings'. The movement and transformation of water within these processes as described in the definition, as a fluid, will obey the physical rules of fluid mechanics. Fluid mechanics, being a quantitative topic, requires heavy use of mathematical concepts, and these concepts are therefore naturally found in hydrology. These quite basic physical principles can be used effectively to model and hence predict and understand the behaviour of water under many useful circumstances.

Nonetheless, despite the essentially predictable behaviour of water that justifies the use of mathematical principles, often, the flow of water in practice is subject to forces that are beyond our ability to measure with any precision: for example, water in the atmosphere is heated, cooled, mixed with numerous gasses, and transported across large distances under the action of turbulent winds. Eventually, water condenses out of the atmosphere in the form of precipitation but exactly when, where, and how much water falls to the ground under gravity is often extremely uncertain. This uncertainty usually makes it useless to apply the basic physical principles of fluid mechanics to the flow of water in these circumstances. For this reason, hydrologists often turn to statistics, which can be considered as the application of mathematics to uncertain phenomena.

Quantitative hydrology is, therefore, based on an interesting mix of the two great branches of applied mathematics: physical laws (mathematical physics) and probability (mathematical statistics).

Mathematics is, perhaps, the archetypal example of a composite subject. This means that more complex concepts are built from many simpler ones, and so, in order to properly understand the more complex topic, it is necessary to understand the simpler ones from which it is constructed. Not all subjects are like this: it is possible to gain a deep understanding of many aspects of plant biology without

Understanding Mathematical and Statistical Techniques in Hydrology: An Examples-Based Approach, First Edition. Harvey J. E. Rodda and Max A. Little.
© 2015 Harvey J. E. Rodda and Max A. Little. Published 2015 by John Wiley & Sons, Ltd.

having to know anything about mammals, for instance. But mathematics is unforgiving: one cannot understand the true meaning of equations of fluid transport without knowing calculus. Unfortunately, for many reasons, the chance to learn the basic mathematical concepts is not afforded to every student or practitioner of hydrology, and many find themselves at a loss when presented with more complex mathematical concepts as a result.

This book is therefore, intended as a guide to students and practitioners of hydrology without a formal or substantive background in either mathematical physics, or mathematical statistics, who need to gain a more thorough grounding of these mathematical techniques in practical hydrological applications.

1.2 Mathematical preliminaries

This book refers extensively to many essential, but nonetheless quite simple, mathematical concepts; we introduce them here. It is assumed that readers will refer back to this section on reading the later material.

1.2.1 Numbers and operations

Usually when one thinks of 'mathematics', one thinks of numbers, along with operations such as adding, subtracting, multiplying (forming the *product*) and dividing them. Numbers and operations are intimately related: for example, with the simplest of numbers, the *whole numbers*, we can answer questions such as 'what number, when added to 5, gives 10?' Symbolically, we wish to find the x that satisfies the equation $x + 5 = 10$, the answer being $x = 5$. But some simple questions involving whole numbers cannot be answered using whole numbers, for instance, the problem 'what number, when added to 10, gives 5?', or $x + 10 = 5$, has no whole number answer. To solve such a problem, we need to include *negative numbers* and *zero*; mathematicians call these whole numbers that can be negative, zero, or positive, the *integers* (all the positive whole numbers are included in the integers). Still, when faced with whole number problems involving multiplication, integers may not suffice. For example, the problem 'what number, when multiplied by 5, equals 1?', or $5x = 1$, has no integer solution. The answer $x = 1/5$ is called a *rational number* and all the integers are included in the rationals. Finally, it turns out that there are yet more problems involving multiplication that cannot be solved using rationals; consider the problem 'what number, when multiplied by itself, equals 2?' The corresponding equation $x \times x = 2$ is solved by the *square root* of 2, $x = \sqrt{2}$, which is an example of a *real number*. The set of real numbers includes all the rationals and numbers such as $\pi = 3.14159\ldots$ (which can never be written out to full precision because it has an infinite number of decimal places). With the set of all real numbers, a very large set of problems involving numbers and operations that do actually have a solution can be answered.

It is surprising that even in apparently simple situations such as multiplication and addition with whole numbers, that there are equations that have no solution in the rationals, let alone the integers and whole numbers. Such equations baffled mathematicians until the 19th century when a logically consistent foundation for the real number system was devised. But real numbers do not even suffice for all whole number equations! Consider the equation $x \times x + 2 = 0$; because squaring any number is always positive, there does not seem to be any way to choose a number for x that, when squared, gives a negative number to cancel the 2 and satisfy the equation. Nevertheless, it turns that a consistent solution is possible using *complex numbers*; although abstract, these can be useful in physical problems.

These days, because of their practical utility, real numbers tend to be the lifeblood of quantitative sciences including hydrology. For instance, the average amount of rainfall occurring in one day in one location is often given as a real number in millimetres, to a couple of decimal places where such precision is appropriate. Therefore, most practical problems in hydrology involve solutions that are real numbers given to some limited accuracy appropriate to the problem.

1.2.2 Algebra: rearranging expressions and equations

An important step in the historical development of mathematics was the leap from dealing with specific numbers, to dealing with *any* number by using an abstract symbol to stand for that number (this conceptual leap is usually credited to the great Islamic mathematicians of the medieval period). This is the topic of *algebra*: the study of what happens to these symbols as they are manipulated as if they were numbers. Most quantitative problems in the physical sciences can be expressed and solved algebraically.

Algebra involves very simple rules. Although the rules themselves are elementary, the consequences of those rules can be extremely complex; in fact, much research still goes on today to understand the full, logical consequences of algebra. For this reason, one should not underestimate how difficult it can be to correctly derive the consequences of any particular application of algebra in practice, and it is very much worth the effort to become as familiar as possible with the basic rules.

Today, one usually writes something like x or y when one wants to refer to an abstract number; these are also called *variables* (as opposed to specific numbers, which are *constants*). Then the notation $x + y + 1$ is an *algebraic expression* using these two variables and the constant 1.

Expressions on their own do not 'do' anything; to make expressions useful we need to connect them together into e*quations*, for instance, the equation $x + y + 1 = 0$ states that if the variables x and y are added to the constant 1, then the result must be equal to zero. Alternatively, by manipulating (*rearranging*) this equation, we can get the exactly equivalent statement $x + y = -1$, which is obtained by subtracting 1 from both sides. This is an example of a basic rule in algebra: in order to rearrange an equation, one has to apply the same operation to both sides of an equation, step by step. This rule ensures that before and after the manipulation, the equation still has the same mathematical meaning.

These *algebraic operations* come in pairs – subtraction is the inverse of addition and division is the inverse of multiplication. What this means, roughly, is that subtraction 'undoes' addition and division 'undoes' multiplication. So, actually, what one is doing when rearranging an equation, is applying a sequence of inverse operation to both sides of an equation.

Rearranging equations is fundamental to the way in which answers to mathematical questions are obtained, often by finding the actual number (value) of some variable. For the equation $x + y = -1$, we only know the value of x and y *implicitly* (through the relationship created between them by the equality). However, it is often difficult (if not impossible) to find the value of x from an implicit equation. In this case, the solution is easy of course: rearrange the equation to find x alone on one side of the equation, for instance, $x = -y - 1$ (note that it does not matter on which side x appears). Then, we can usually find a unique value for x, because the right-hand side of the equation is an *explicit* formula for *solving for* the value of x.

The 'art' of rearranging equations to solve for a particular variable, then, is to find a sequence of steps that can be applied to both sides of the equation such that we end up with that variable alone on one side of the equation. Unfortunately there is no general procedure for the 'correct' sequence of steps to apply to any equation: efficient equation solving is often a matter of experience and practice.

The operations of addition and multiplication have the important property that when applied to two or more variables or constants (*terms*), the order in which they are applied does not matter. For instance, for the product $2 \times x \times y = 2 \times y \times x = x \times 2 \times y$, etc. The same applies if we replace the product with addition: $2 + x + y = 2 + y + x = x + 2 + y$. *But* when *combining* different algebraic operations, the order in which variables, constants and operations appear in an expression is critical. For example, $2 \times x + y$ is not the same as $2 \times (x + y)$. The brackets in the second equation indicate that first, x should be added to y, and then the result should be multiplied by 2. In fact, by *expanding* the brackets, the second expression becomes $2 \times x + 2 \times y$, which makes it clear that it is not the same as $2 \times x + y$. A tricky example of this is the expression $x - y$: this is *not* equal to $y - x$. In fact, $x - y = -y + x$. The reason is that actually, the expression $-y$ is shorthand for $(-1) \times y$, and we have to take account of the fact that the multiplication of y by -1 must happen before the addition to x.

The general advice then about rearranging more complex expressions and equations is that carefully and systematically, examine the *order* in which the algebraic operations are supposed to be applied to the terms. Modern algebraic notation has some conventions for this (called *precedence* rules); unless otherwise overridden using brackets, multiplication and division occur before addition and subtraction. As a case in point, consider the following expression:

$$\frac{x+y}{3y} \tag{1.1}$$

This could be interpreted as follows: first add x to y, then multiply y by 3, and then divide the first result by the second result. It *does not* say, for example, multiply y by 3, divide y by this, and then add x; the following is an example of such kind:

$$x + \frac{y}{3y} \tag{1.2}$$

Another way of explaining the difference is that (1.1) can also be written using brackets as $(x+y)/(3y)$ – then the ordering becomes clear. In (1.1) and (1.2), we can apply some rearrangements that might be useful, for example, by expanding out the 'brackets' in (1.1), we get that $\frac{x+y}{3y} = \frac{x}{3y} + \frac{y}{3y}$ (applying the rule that dividing by some expression is equivalent to multiplying by 1 divided by that whole expression). Next, we can apply the rule that dividing an expression by itself is equal to 1 (unless that expression is equal to zero – see below); so $\frac{x+y}{3y} = \frac{x}{3y} + \frac{1}{3}$. For (1.2), we get $x + \frac{y}{3y} = x + \frac{1}{3}$ for the same reason. Slightly more complex is the situation where the top and bottom expressions both involve addition, for instance, as follows:

$$\frac{x+y}{3x+3y} \tag{1.3}$$

To rearrange this expression, we consider *factoring* the bottom part of the division as a rearrangement step. A *factor* is a number (or variable) that multiplies another expression, for instance, the expression $6xy$ has the factors 6, x and y (actually, since $6 = 2 \times 3$, it is also reasonable to argue that there are four factors 2, 3, x and y). The factored bottom expression is then $3x + 3y = 3(x+y)$. Effectively, we have changed the order of the multiplication by 3 and the addition: this is what factoring achieves. So, factoring undoes expanding out brackets. Now it is clear to see that if $x+y$ is not zero, (1.3) has the value $1/3$, that is the $x+y$ expression cancels completely.

The number zero has a special importance in algebra. Firstly, note that adding zero to some expression leaves that expression unchanged, for example $x + 0 = 0 + x = x$ (An interesting observation is that 1 plays the same role in multiplication as 0 takes in addition, namely, it leaves the expression unchanged: $x \times 1 = 1x = x$.) The second property of zero is that multiplying some expression by zero results in zero, for example $0x = x \times 0 = 0$. Sometimes, we end up with an equation such as $0 = xy$. Using the second property, we can see that one or more of x and y must be zero for this equation to be satisfied. Another consequence of these properties is that dividing anything by zero is undefined (effectively, there is no meaningful result). Consider what it means to write $x = y/0$. For the moment, treating 0 as a symbol, we could rearrange this equation to $0x = y$, and applying the second property of zero y must be zero. However, then the equation becomes $0x = 0$, and this is true for *any* value of x! In other words, the original equation, even though it is an explicit formula for x, does not tell us what value x should take. Because of this, (1.1) and (1.2) are meaningless in the special case where $y = 0$ and (1.3) is meaningless if $x + y = 0$.

Often, we have the situation where there are two or more variables whose value we need to find in order to solve a practical problem. In general, we need as many equations as there are variables in order to find a solution that gives the unique values of all the variables. For instance, to solve the following pair of equations for x and y,

$$x + y = 0 \tag{1.4a}$$

$$3x + y = 1 \tag{1.4b}$$

We might want to solve for x in (1.4a), $x = -y$, and then *substitute* this expression for x into (1.4b), $3(-y) + y = 1$. We can then factor out y in this to obtain $(-3 + 1)y = 1$, and we get an explicit formula $y = 1/(-2) = -1/2$. Now, using the explicit formula for x, it must be that $x = 1/2$, and we have a solution for both variables. This is a simple example that illustrates how to apply sequential rearrangement and substitution in order to solve a pair of equations; this basic principle can be attempted for more complex equations but it usually becomes very difficult in practice to solve equations involving three or more variables. Typically, one then turns to *computer algebra* software, or, instead uses *numerical methods* to obtain approximate solutions.

Repeated self-multiplication of some term or expression has a special name: *exponentiation* ('raising to the power') and is written using the superscript notation as x^n, where n is called the *power* or *exponent*. If n is a whole number, this just means that we multiply x by itself n times. So, that means that $x^1 = x$. If we have powers n and m that are both whole numbers, then it is fairly easy to see that $x^n \times x^m = x^{n+m}$. In a sense, we can see that this rule 'converts' multiplication into addition. When applied to expressions, there are some simple consequences, for instance when $n = 2$, the expression $(x + y)^2 = x^2 + 2xy + y^2$ (which one can check by expanding out the brackets – there is a general formula called the *binomial expansion* that works for any general value of n).

If we allow $n = 0$, then $x^0 \times x^m = x^{0+m} = x^m = 1 \times x^m$, so it is reasonable to claim that $x^0 = 1$ (to be consistent with the role that 1 plays in multiplication, as discussed above). Similarly, if we allow that $n = -m$, then we get $x^{-m} \times x^m = x^{m-m} = x^0 = 1$, so we can claim that $x^{-m} = 1/(x^m)$ to be consistent with the idea that dividing some expression by itself is equal to 1. It follows then that $x^{-1} = 1/x$. In fact, it can be shown that this rule converting multiplication into addition works quite generally: n and m can be any fraction or real number and thus exponentiation is a general algebraic operation. Certain rational powers are given special names: $x^{1/2}$ is called the square root and written $\sqrt{x}$, and more generally, $x^{1/n}$ is the nth *root* written $\sqrt[n]{x}$.

Being a general algebraic operation, exponentiation has an inverse called the *logarithm* (a key mathematical discovery credited to John Napier in the 16th century). We write this as $\log_x y$, where x is called the *base* of the logarithm, which has the meaning that if $n = \log_x y$, then $x^n = y$: the logarithm to base x recovers the power of x (so, for example, $\log_x \sqrt{x} = 1/2$). As we demonstrated above, since exponentiation 'converts' multiplication into addition, we can explain that the logarithm converts addition back to multiplication. Consider the power rule $x^n \times x^m = x^{n+m}$, then taking the logarithm to base x on both sides, we get $\log_x(x^n \times x^m) = \log_x(x^{n+m}) = n + m = \log_x x^n + \log_x x^m$. As with exponentiation, this rule actually works for general numbers, not just whole numbers, and we can derive some consequences worth memorizing: $\log_x 1 = 0$ (which is the inverse of $x^0 = 1$), $\log_x x = 1$ (which is the inverse of $x^1 = x$), $\log_x x^n = n$, and the general rule $\log_x(a^n \times b^m) = n\log_x a + m\log_x b$ for any numbers a, b, n, m and x, provided only that neither a nor b is zero. The last rule has a useful special case: $\log_x(a/b) = \log_x(a \times b^{-1}) = \log_x a - \log_x b$.

In practice, since logarithms in one base can be converted to any other base using the formula $\log_x a = \log_y a / \log_y x$, one tends to work in a standard base such as 10. The other commonly used base is the *natural logarithm* which uses the base $e = 2.71828\ldots$ (we will see later that this has a very important origin), written as $\ln x$. The inverse to the natural logarithm, $e^x = \exp(x)$, plays a very important role in much of mathematics: as the inverse it follows that $\ln(\exp(x)) = x$.

1.2.3 Functions

Expressions are very often 'packaged up' into convenient shorthand notation known as *functions*, such as $f(x)$ or $g(x)$. Examples of functions include the exponential function $\exp(x) = e^x$ and $\ln(x)$ above but also familiar functions such as the trigonometric functions $\sin(x)$, $\cos(x)$ and $\tan(x)$. Use of functions in expressions can improve the readability of equations considerably. Very often there is an associated *inverse function*: as we have seen, $\exp(x)$ has $\ln(x)$ as its inverse. Sometimes, consideration of the range of acceptable values that a function can take tells us about the range of output values of its inverse: for example, the $\sin(x)$ function takes all possible real *angles* as input, but its output is restricted to the range −1 to 1. So, the inverse function $\sin^{-1}x$ can only *accept* numbers in the range −1 to 1.

Plotting a function as a *graph* can be very useful; typically this is done by drawing a curve on axes where x is on the horizontal and $y = f(x)$ is on the vertical. Then, since a function only outputs one value per unique input value, the curve must be a single, non-self-intersecting line. In addition, very often that line can be drawn without taking the 'pen' off the paper, so the function has no *discontinuities*. Functions can take more than one number as input, for example $f(x,y) = x^2 + 3y^3$; this makes it much harder to plot a graph of the function (which would appear as a surface in 3D with $z = f(x,y)$ being the height of the surface).

1.2.4 Calculus

The name given to the theory of mathematics that deals with the abstract concepts of area (including length and volume more generally) and gradient (slope) is called *calculus*. Although the mathematicians of the ancient world knew how to calculate these quantities for simple shapes (for example, working out how to divide up a rectangular field into equal areas for the purpose of probate law), they did not know how to do this for general geometric objects, particularly if they had arbitrarily curved boundaries. This had to wait until the 17th century for the mathematical innovations of Newton and Leibniz, who saw the potential for applying and extending these concepts to predicting the motion of the planets. It is probably fair to say that the vast majority of physical applied maths revolves around the use of concepts from calculus.

Summation plays a central role in calculus: we write $\sum_{i=0}^{N} x_i$ to denote the sum from 0 to N of the values in the $N+1$ variables x_i (we use the subscript notation to index each of these different variables). We can apply this to the problem of calculating areas – the area of a rectangle is just $A = w \times h$ where w is the width and h is the height. Now, if a given shape can be approximately broken down into $N+1$ small rectangles, then the area of the complete shape is approximately as follows:

$$A \approx \sum_{i=0}^{N} w_i h_i \tag{1.5}$$

In words, 'the sum of the product of the width of each rectangle times the height of rectangle is approximately the total area of the shape'.

If we can assume that the width of all these rectangles is the same, we can simplify this to $\sum_{i=0}^{N} w h_i$. Of course, this will only be an approximation to the area, for example some of the area might not be counted.

Assuming, for the sake of simplicity, that one edge of the object is straight and lies on the x-axis of graph and the other side is represented by a function with arbitrary curves (this requirement might seem contrived but it turns out that calculus can be defined in more flexible ways for different geometric situations, using essentially the same ideas). Figure 1.1 shows this idea for finding the area (*integrating*) under the curve $f(x) = x^2$. In the upper panel, we have a relatively coarse set of rectangles with equal width w attempting to fill the area; the bottom panel has a much slimmer set of rectangles, again of fixed width. It is easy to see that the amount of uncounted area in the bottom panel is smaller than that in the top panel, so the bottom panel is a better approximation to the area. In calculus,

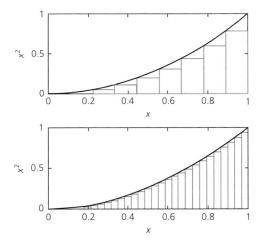

Figure 1.1 Approximate integration of the area under the curve x^2 (black) using rectangles (grey), over the interval 0–1, with coarse partition (top) and finer partition (bottom).

the idea is to calculate what happens as the width of the rectangles becomes arbitrarily small, following from the intuition that slimmer rectangles give better approximations: of course, the number of rectangles will become arbitrarily large as a result. The approach aims to convert the problem of finding the area under the curve to a problem of finding the ultimate value of a sequence of better and better approximations.

To do this, we will need the idea of *limits*, one of the core concepts of calculus. Mathematicians use the shorthand notation '$a = \lim_{x \to c} f(x)$' for the limiting value of the function $a = f(x)$ as x takes on values that are always getting closer to c. This is also written as '$f(x) \to a$ as $x \to c$'. A critical point to understand is that in most useful cases, the limiting value a cannot be calculated directly. For example, it makes intuitive sense (and it is logically correct as we will show next) that $\lim_{x \to \infty} 1/x = 0$. But since infinity does not have a *definite* value, algebraic expressions such as $1/\infty$ do not have a definite result either.

Limits are a (indirect) way of computing definite answers in these situations. For example, we know that the function $f(x) = 1/x$ is continuous (see above) at all values of x except 0. Also, the function is *decreasing*, that is if we pick any two numbers x and y such that $x < y$, then $1/x > 1/y$. Additionally, we know that if x is positive, then $1/x$ is also positive. These pieces of information allow us to conclude that $1/x \to 0$ as $x \to \infty$. In other words, we have shown that as we keep increasing the (positive) value of x, $1/x$ always gets smaller, and since it cannot be negative, as x becomes arbitrarily large (infinite), $1/x$ must ultimately take on the value zero. At root, this is typical of limit value arguments: nonetheless, most problems encountered in practice are reducible to an algebraic combination of known results about the limits of basic functions.

We now return to the problem of finding the limit of sequence of approximations to the area under the curve. We can construct a grid of x-values as $x_i = wi$ and the corresponding height of the rectangles is $h_i = (wi)^2$. Then the number of rectangles in the interval 0–1 is $N = 1/w$. So, the area is written as:

$$A = \lim_{w \to 0} \sum_{i=0}^{N} w(wi)^2 \qquad (1.6)$$

This equation states that first, we sum up all the areas of the rectangles fitting underneath the curve. Then, we take the limit of these sums, as the width of these rectangles becomes arbitrarily small. For every rectangle width, there will be a corresponding number of rectangles N, which therefore must go to infinity as the width goes to zero.

Here we make the remark that in this specific case (1.6) does have an exact answer, $A = 1/3$, that we can compute using well-known, but somewhat complex, algebraic manipulations.

If we want to compute the area under an arbitrary function $f(x)$ over any chosen range of values of the x-axis, say, from a to b, we need the *definite integral*:

$$\int_{a}^{b} f(x)dx = \lim_{N \to \infty} \sum_{i=0}^{N} f(a+iw)w \qquad (1.7)$$

where we choose $N = (b-a)/w$ (note the w is often written as Δx, because w is a 'small difference in x'). In Equation (1.7), the left-hand side is just shorthand for the right-hand side, which states that the area is computed by summing up rectangles of width w, placed at each position on a grid of spacing w covering from a and b on the x-axis. The particular choice of N means that when $i = N$, $a + iw = b$, the right-most grid position. Each rectangle has height $f(a+iw)$.

Note this is only a definition: there is no guarantee that we can actually find the limit of the sequence of approximations to find the exact answer by some straight-forward algebra. In fact, the somewhat disappointing news is that the number of functions $f(x)$ that we *cannot* integrate in this way vastly outnumbers the functions that can be integrated like (1.6). This usually happens because the kind of algebraic tricks used to remove the summation in situations such as (1.6) work only in special cases. Nonetheless, under certain conditions that are not too restrictive, we can say that the limit in (1.7) is useful, in that, it has a definite value, and we can approximate this to any desired accuracy using a computer program, for example.

The inverse operation to integration is *differentiation*. It is relatively simple to find the gradient of a straight line. Imagine finding the slope of a straight road running up a hill with constant angle to the horizontal: it is just the *rise over the run* or the change in vertical height (Δy) you go through as you travel over some horizontal distance (Δx):

$$m = \frac{\Delta y}{\Delta x} \tag{1.8}$$

This is how to calculate the gradient of a function if it is a straight line. How can we do this if the function is not a straight line? One way is to assume that over small enough distances, the slope of *any* function at a particular fixed point can be approximated by the slope of a straight line that goes through that point. This will be a good assumption if the function is smooth enough. As the distance over which we make this assumption gets smaller, the approximation to the slope at that point gets better. For a given function $f(x)$, the change in 'height' at x over the distance Δx is $f(x + \Delta x) - f(x)$, and therefore, using the idea of limits to define the derivative, we get:

$$m(x) = \lim_{\Delta x \to 0} \frac{\Delta y}{\Delta x} = \lim_{\Delta x \to 0} \frac{f(x + \Delta x) - f(x)}{\Delta x} \tag{1.9}$$

(Note that the slope of a general function is itself a function of x, the chosen point, unlike a straight line, which has the same slope at every point.) In this way, *differentiation* solves the problem of how to find the slope of a function which is arbitrarily curvy. The derivative is also commonly written as df/dx or also $f'(x)$ when it is clear that we are differentiating with respect to x.

Algebra that arises from differentiating is usually a lot simpler than algebra that arises as a result of integration. For this reason, many more functions can be algebraically differentiated than integrated. As with integration, under not very restrictive conditions, the limit in (1.9) is useful and can be calculated approximately to any desired precision numerically.

There is a theorem in calculus that relates integration and differentiation (called, appropriately enough, the 'fundamental theorem of calculus'). This theorem can be stated in many different ways, but it is instructive to provide geometric intuition. Firstly, consider the area under the function $f(x)$ from 0 to x written as $F(x)$. Now the area under the curve between x and $x + \Delta x$ can be found as $F(x + \Delta x) - F(x)$, which is the area from 0 to $x + \Delta x$ minus the area from 0 to x. However, as above, when defining the integral, we could also approximate the area between x and $x + \Delta x$ with the small rectangle of area $f(x)\Delta x$. It follows that $F(x + \Delta x) - F(x) \approx f(x)\Delta x$ or

$$f(x) \approx \frac{F(x + \Delta x) - F(x)}{\Delta x} \tag{1.10}$$

Taking the limit of both sides as $\Delta x \to 0$ gives us $f(x) = F'(x)$, using the definition of the derivative above. So, this says that the original function $f(x)$ is what we get by finding the slope of the area under the curve of that function.

The following is the most common statement of the fundamental theorem:

$$\int_a^b f(x)dx = F(b) - F(a) \tag{1.11}$$

This says that the definite integral of the function $f(x)$ is the area under the function from 0 to b minus the area under the function from 0 to a. This allows us to introduce the *indefinite integral*, useful when integrating from 0 to the value of some variable x:

$$\int f(x)dx = \int_0^x f(x')dx' = F(x) - F(0) = F(x) + c \tag{1.12}$$

(Note that we had to use a *dummy variable x'* to avoid a conflict between variable names, because x normally does not appear in the integration range). The replacement of $-F(0)$ with the generic constant c indicates that an arbitrary constant is always introduced when performing indefinite integration.

Yet another statement of the fundamental theorem is:

$$\int_a^b f'(x)dx = f(b) - f(a) \tag{1.13}$$

We get this from (1.11) by replacing the function $f(x)$ with the derivative of the function $f'(x)$ instead. This form of the theorem tells us something quite profound about calculus that has far-reaching consequences in many areas of mathematics: the definite integral of the slope of the function is just the difference of the function value at the far end of the range, minus the function value at the near end.

The derivative and integral have certain important algebraic properties of their own that are consequences of the way they are defined. The first is the fact that they are both *linear operations*:

$$\int [af(x) + bg(x)]dx = a\int f(x)dx + b\int g(x)dx \tag{1.14a}$$

$$\frac{d}{dx}[af(x) + bg(x)] = a\frac{df}{dx}(x) + b\frac{dg}{dx}(x) \tag{1.14b}$$

In the above, $f(x)$ and $g(x)$ are arbitrary functions, and a, b are constants, and the equations state that it is possible to swap the order of scaling, addition and integration, and the same with differentiation. This means that we can first multiply two functions by constants, add the results together and then integrate, or we can first integrate the functions separately, multiply the results by constants and then add the results together. The same applies to differentiation. It is critical to note that the above rules only apply if a, b do not change as x changes.

More algebraic properties apply to differentiation, for example, the *product rule* states what happens when differentiating the product of two functions:

$$\frac{d}{dx}f(x)g(x) = f(x)\frac{dg}{dx}(x) + g(x)\frac{df}{dx}(x) \tag{1.15}$$

Similarly, the *chain rule* explains what happens if we differentiate a function of a function:

$$\frac{d}{dx}f(g(x)) = \frac{df}{dx}(g(x))\frac{dg}{dx}(x) \tag{1.16}$$

There are further 'rules' that occur in more complex combinations. It is important to grasp that (1.14a) is the *only* genuine algebraic property of differentiation shared by integration – properties (1.15), (1.16) have no direct counterparts in integration. There are other integration 'rules', such as the integration by parts and integration by substitution, but these are actually obtained by 'undoing' the rules of differentiation (1.15) and (1.16).

Differentiating a function twice gives the *second derivative* (also known as the *curvature* of a function):

$$\frac{d}{dx}\left(\frac{d}{dx}f(x)\right) = \frac{d^2}{dx^2}f(x) = f''(x) \tag{1.17}$$

Similarly, the *n*th *derivative* is written as d^n/dx^n, for $n > 0$, or sometimes $f^{(n)}(x)$.

It is helpful to list a few specific derivatives and integrals. Perhaps the most important is the exponential function encountered earlier:

$$\frac{d}{dx}\exp(x) = \exp(x) \tag{1.18a}$$

$$\int \exp(x)dx = \exp(x) + c \tag{1.18b}$$

This explains one important reason behind the special place of the exponential function in mathematics as it is the only function that is simultaneously its own integral and derivative. Other important functions include the powers of *x* (*polynomials in x*):

$$\frac{d}{dx}x^n = nx^{n-1} \tag{1.19a}$$

$$\int x^n dx = \frac{1}{(n+1)}x^{n+1} + c \tag{1.19b}$$

Many other functions are explicitly differentiable and integrable in this way, notably the trigonometric functions (sine, cosine, tangent), but most functions do not have simple integrals and we usually turn to numerical algorithms to compute them for particular ranges in practice.

1.2.5 Differential equations

Having defined differentiation as the slope of a function with arbitrary 'curviness', since the slope depends on the chosen value of x, the derivative of a function is a new function of x. So, as with any function, it can usefully appear in an equation alongside other functions, operations, and constants. The resulting *differential equations* have been the cornerstone of modern physical applied mathematics ever since Newton's *Principia*: the sheer number of physical problems that can be formalized using differential equations is truly staggering.

As an example, consider the following differential equation:

$$\frac{d}{dx}f(x) = m \tag{1.20}$$

where m is just a constant that we know. This states that the equation is satisfied if $f(x)$, when differentiated, is constant. A moment's thought will lead to the answer that $f(x)$ must describe a straight line on the graph of the function: there is no other function whose slope is always the same constant. We can also prove this by integrating both sides of (1.20), finding that $f(x) = mx + c$, which is indeed an expression for a line on the graph of x against $f(x)$, called the *general solution* to the differential Equation (1.20). We can check that we have the right solution by differentiating $f(x)$, and testing that this satisfies the equation. One important point to note is that an arbitrary constant c appears in the solution (because we are performing indefinite integration) so, without specifying this constant, we cannot find the value of the solution for a given x. In this case, the constant can be set by specifying an *initial condition*: that is what we expect $f(x)$ to be when $x = 0$. For instance, the initial condition $f(0) = -2$ leads to the *specific solution* $f(x) = mx - 2$.

Perhaps the most famous of all elementary differential equations is the simplified model of the mass on a spring, ignoring friction:

$$mf''(t) = -kf(t) \tag{1.21}$$

where m is the mass, and k is the spring stiffness, and t represents time. In physics, the quantities $f(t)$, $f'(t)$ and $f''(t)$ have special names: *position*, *velocity*, and *acceleration* (those with a physics background might recognise (1.21) as an application of *Newton's second law*). So, (1.21) states that the acceleration, multiplied by the mass, is equal to the negative of the position multiplied by the spring constant. We will also assume that the position at time zero is some constant A: $f(0) = A$, and the initial velocity is zero: $f'(0) = 0$. Special algebraic techniques have been developed to find solutions to equations such as (1.21), when applied, the specific solution is:

$$f(t) = A \cos\left(\sqrt{\frac{k}{m}}t\right) \tag{1.22}$$

The cosine arises because it can be shown that if $f(x) = \cos(x)$, then $f''(x) = -\cos(x)$, that is the cosine function is the negative of its' own second derivative, which is, essentially, what is required to satisfy (1.21). The character of this solution is *oscillatory*: that is, the mass vibrates at a rate of $\sqrt{k/m}$, with amplitude A. So, if the mass is increased, the vibration becomes slower, and if the spring stiffness is increased, the vibration speeds up (an intuitive result). Note that the rate of vibration is not dependent upon the initial position A.

Differential equations used in quantitative hydrology can be a lot more complex than (1.21), but the principles are the same. Most of the complexity arises when dealing with functions of more than one variable. For example, a function $Q(x, t)$ might represent the quantity of water in a channel in both time and position. Then, we need to introduce *partial differential equations* that involve the derivative of the function in one or more variable at a time, for example, the (one-way) *kinematic wave equation* in hydrological flow routing is:

$$c\frac{\partial Q}{\partial x}(x,t) = -\frac{\partial Q}{\partial t}(x,t) \tag{1.23}$$

The notation $\partial Q/\partial x$ is shorthand for the derivative of the function $Q(x, t)$ with respect to x alone:

$$\frac{\partial Q}{\partial x}(x,t) = \lim_{\Delta x \to 0} \frac{Q(x+\Delta x, t) - Q(x,t)}{\Delta x} \tag{1.24}$$

Equation (1.23) states that the rate of change of the quantity in space, multiplied by c is the negative of the rate of change of the quantity in time. Again, techniques have been developed to solve such equations to find an explicit expression for $Q(x, t)$. This equation has some similarities to (1.21): except that it involves only first derivatives, and two variables instead of one.

Many of the equations of quantitative hydrology are partial differential equations such as (1.23). A large number of useful ones (such as the *shallow water wave equation*) are, unfortunately, unsolvable using the kind of algebraic tricks that are available for solving (1.21) and (1.23). For this reason, computational algorithms involving purely numerical calculations have been devised and are an important tool in modern quantitative hydrology.

1.2.6 Probability and statistics

Statistical techniques form a critical part of the material in Chapters 2, 6 and 7. Statistics is based on the mathematics of uncertainty, known as *probability*. By comparison to the other areas of mathematics covered above, probability as a mathematical topic is a relative newcomer, having origins in the 17th century. As with algebra, the basic rules of probability are elementary and intuitive: but the logical consequences of these rules, particularly when applied to real-world data, can be complex and often counter-intuitive. For instance, consider the chance of rain falling on the city of Oxford in the UK in any one day. The (extensive) historical data suggests that it is as likely to rain on any day as not, and that whether it rained yesterday or the day before has almost no influence on whether it rains today. So it is quite accurate to model this situation as the flipping of an unbiased coin: the probability of rainfall in one day is 1/2, and previous coin flips do not influence the current one. When this is explained to people, most will surmise, correctly (presumably from experience), that the probability of there being 30 rain days in a row is extremely small. Because of this, if, in the unlikely event that 30 days of rain did actually occur, many will assume that the next 30 days will have to be dry in order to maintain the expected probability of 1/2. But this belief is false: according to the coin-flipping model, whether it rained yesterday or on any previous day has no bearing on whether it rains today. We have simply witnessed an extraordinarily unusual event. Of course, if such an event *did* occur in the historical record, it would distort the statistics so much that we might decide that the unbiased coin model is not actually appropriate.

The mathematical ingredients of probability are easy to state. There is the set of all possible *outcomes* relevant to the physical situation. For example, the cloudiness at any one time in one location can be observed as clear, scattered clouds (~25%), partly cloudy (50% coverage), mostly cloudy (75% covered) or overcast; or it can rain or not on any one day at a specific location; or the rainfall in any one day in one location can be zero or more millimetres, in steps of one-tenth of a millimetre. From these outcomes, we form *events* of interest, the probability of which we want to know. For example, the event that the rainfall depth is greater than or equal to 10.0 mm, or the event that there are scattered clouds.

To these basic ingredients are added three rules (known as *axioms* by mathematicians):

Rule 1. To each possible event is attached a real number called the *probability value* that must not be negative;

Rule 2. Since one of the outcomes is certain to occur, the event that any one of the outcomes is observed is assigned the probability value 1. From this, and the previous condition, we can conclude that probabilities of events lie between 0 and 1 inclusive, with 0 meaning *impossible*, and 1 denoting *inevitable*;

Rule 3. The probability of any compound event obtained by joining mutually exclusive (that is, non-overlapping) events together, is just the sum of the probabilities of the individual events.

Some examples are useful. Consider the event that rainfall is at least 10.0 mm, and assume that the probability of this event is known to be 0.01. The *complementary* event is that the rainfall is less than 10.0 mm. These two events are mutually exclusive because any outcome (rainfall depth in 0.1 mm steps) satisfies one of the events, but not both simultaneously. In fact, these two events are also *exhaustive* because joined together, they cover any outcome. The probability of the 'joint' event, which is just 'some rainfall depth occurs' is 1 (from rule 2). This means that the probability of the complement event (rainfall depth is <10.0) must be $1 - 0.01 = 0.99$ (to satisfy rule 3).

As another simple example: if we know that any of the five categories of cloud cover are equally likely, then the probability of any category is just 1/5. Then, since it cannot be both clear and overcast at the same time, the probability of the location being overcast or clear is $1/5 + 1/5 = 2/5$ (using rule 3).

Usually, when dealing with random quantities in physical problems, there is an intuitive numerical label for each possible outcome. If this label is a whole number or integer, then the random quantity X can be associated with a *probability mass function P* (PMF – also known as a *distribution*), that determines the probability value assigned to each event, which lies between 0 and 1. From the rules above, it must be the case that the sum of the distribution of each outcome must be 1. For example, if the outcomes are labelled as integers from zero and above, then $\sum_{i=0}^{\infty} P(X = i) = 1$ to satisfy rule 2.

When the random quantity is a real number, say, depth of a river, then the quantity X is associated with a *probability density function p* (PDF – although it is common to call this a distribution as well). With real-valued variables, some subtleties occur. Firstly, the PDF itself does not represent a probability value: we need to invoke calculus to find the *area under the PDF* which gives the probability. For example, the probability that X lies between 2 and 3, for instance, is $\int_2^3 p(x)dx$. From the property of the integral that $\int_a^a p(x)dx = 0$, we can infer that the probability that X takes on some single value a is always actually zero. Also, the probability density function can be larger than 1: this does not violate rule 2, because the area under the PDF must sum to 1, that is $\int p(x)dx = 1$. Therefore, the area under any smaller range of values than all possible values of X will be less than 1.

Statistical hydrology is often interested in calculating the probability of some event based on a *probability density model* for that variable. Many of the models have *parameters*: constants that affect the shape of the density function. Somehow, these parameters have to *estimated* for a particular data record. Parameter estimation is one of the main topics of statistics, which has led to a large range of techniques. Perhaps the most widely-used technique is *maximum likelihood*, which proposes that the optimum parameter values are those that maximize the probability density given the data record.

Particular quantities of distributions have special significance in probability and statistics. The *mean* (often called the *average*) is the *first moment* often written $E[X]$ or $\bar{x}$, and the *variance*, which is the *second (central) moment* written as $E[(X - E[X])^2]$ or $var(X)$. Note that the standard deviation is the square root of the variance.

If we know a model distribution for the random variable X, say, $p(x)$, then these quantities can be calculated using integration:

$$\bar{x} = E[X] = \int x\, p(x)\, dx \tag{1.25}$$

$$\text{var}(X) = E\left[(X - E[X])^2\right] = \int (x - \bar{x})^2\, p(x)\, dx \tag{1.26}$$

We can also estimate these quantities from data. Technically, this is usually done by constructing a distribution based on the data, which places equal weight at each of the N data points $x_1, x_2 \ldots x_N$, and having no density anywhere else. Then, the integrals in Equations (1.25) and (1.26) simplify to:

$$\bar{x} = E[X] = \frac{1}{N}\sum_{i=1}^{N} x_i \tag{1.27}$$

$$\text{var}(X) = E\left[(X - E[X])^2\right] = \frac{1}{N}\sum_{i=1}^{N} (x_i - \bar{x})^2 \tag{1.28}$$

When calculating these values from data, the accuracy of the estimates is very much tied to the amount of data available: however, in most cases it can be shown that the estimated mean improves as the length of the data record improves.

Finally, another important value is the *median*, which is the special value a such that the probability of X being less than a is the same as the probability of it being larger than a: it can be estimated as the value that lies 'in the middle' when all the data is sorted numerically. Similarly, the maximum is the largest value when the data is sorted.

Reference

UNESCO/WMO (1992) *International Glossary of Hydrology*, 413pp. UNESCO/WMO, Paris/Geneva.

CHAPTER 2
Statistical modelling

This chapter considers the statistical and mathematical techniques which are used to quantify the probabilities of hydrological variables such as observed rainfall, flow or drought. Assigning probabilities to such variables invokes *distribution fitting*, a very important and ubiquitous kind of *statistical modelling* activity, in which an appropriate formula is fitted to observed data (*samples* in statistical parlance).

There are very many such formulae and choosing the appropriate one for the data is a complex and often heavily contested affair beyond the scope of this book, but things are very much simplified and constrained when quantifying only the probabilities of *extreme events*, which are often the main focus of interest in practical hydrology. For these extremes, such as a period of heavy rainfall that might have led to flooding, it is most common to refer to the probability of such an event by its *return period*, for example a '1 in 100 year' flood. This idea of return period has often caused confusion amongst those who are not familiar with the term. It does not mean though that the interval between the events will be 100 years. For example, the flood in the Seine at Paris in 1910 was reported by hydrologists as a 1 in 100 year event. This caused great concern in 2010, when the media in France were questioning the hydrological services about being prepared for the imminent arrival of the next severe flood as it was now exactly 100 years since the last one!

The branch of mathematics and associated techniques used to quantify such extreme events is termed *extreme value theory*. Before diving into the maths associated with extreme events, it is worth discussing an example of such extreme occurrences from a hydrological perspective.

2.1 The Central European Floods, August 2002

In August 2002, countries in Central Europe were affected by the worst flooding in living memory. The flooding, in the basins of the Danube and Elbe, affected large parts of Austria, the Czech Republic and Germany and parts of other

Understanding Mathematical and Statistical Techniques in Hydrology: An Examples-Based Approach, First Edition.
Harvey J. E. Rodda and Max A. Little.
© 2015 Harvey J. E. Rodda and Max A. Little. Published 2015 by John Wiley & Sons, Ltd.

countries such as Slovakia and Hungary as the flood waves moved downstream. The flooding was caused by a combination of the antecedent conditions and extreme rainfall from two intense depressions within 6 days of each other which followed an almost identical trajectory.

The preceding months of June and July were uncharacteristically cool and wet which meant river levels were higher than normal and soil moisture deficits were low. The reason for this was that the meteorology over the region was being driven by the jet stream taking a more southerly route across continental Europe rather than over the British Isles as a blocking anticyclone had become established over Scandinavia. This large scale weather pattern, known in meteorological terms as a *Vb circulation*, caused the Atlantic depressions to follow the jet stream over continental Europe. Here the meeting of warm moist air from the Mediterranean with polar maritime air caused uplift, convergence and considerable precipitation. The rainfall totals were enhanced by the slow moving nature of these depressions and the presence of mountain ranges where orographic effects such as funnelling increased convergence and the formation of cap clouds produced record rainfall totals.

The first depression brought rainfall over Austria, Southern Germany and the south of the Czech Republic on 6th and 7th of August 2002, with totals of 200 mm and up to 150 mm in 24 hours. This event caused some significant flooding in smaller rivers and flows of 5–10 year return periods in major rivers. The flows were still high, reservoirs were full to capacity and soils were saturated when the second depression brought more intense rain 5 days later. This time the heaviest rain fell on the western side of the Czech Republic and south-eastern Germany. The Czech–German border region of the Ore Mountains received the highest totals with over 400 mm, a new 24 hour record for Germany (312 mm) was observed at Zinnwald–Georgenfeld and the city of Dresden received a new record 24 hour fall of 158 mm. These values can be put in perspective in relation to the average annual total rainfall for Dresden which is 680 mm.

Extreme flooding ensued following the second rainfall event: initially flash flooding in steep mountainous streams with much damage and bed load transport, and then flooding in the main rivers such as the Vltava at Prague, the Elbe at Dresden and the Danube at Passau. A record flow of 5300 cumecs was measured at Prague, more than 40% higher than the pre-event estimated 1 in 100 year flow of 3700 cumecs (Figure 2.1). Many flood defences were breached and huge extents of the flood plain – up to 15 km from the river channel – were inundated. The flood waves then progressed downstream to areas which were outside rainfall area such as on the Danube in Slovakia and Hungary and the Elbe into central and northern Germany. More than 100 lives were lost, hundreds of thousands of buildings were damaged and the overall economic losses were estimated at 16 billion Euros.

The state hydrological services produced a range of initial estimates of the severity of the flooding based on the return periods at gauging stations

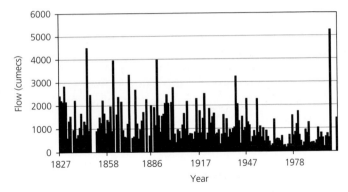

Figure 2.1 Annual maximum flows on the Vltava at Prague, 1827–2006.

Table 2.1 Return period estimates for flows during the August 2002 floods.

Country	River	Location	Date	Peak Level (m)	Peak Flow (cumecs)	Estimated Return Period (years)	Data Source
Czech Republic	Vltava	Ceske Budejovice	13/08/2002	10.00	652	500	CHMI
	Vltava	Praha-Chuchle	14/08/2002	7.85	5300	500	CHMI
	Luznice	Klenovice	15/08/2002	5.80	530	>1000	CHMI
	Otava	Pisek	13/08/2002	8.50	1200	1000	CHMI
	Berounka	Beroun	13/08/2002	7.96	1800	250	CHMI
	Labe	Decin	16/08/2002	12.30	5100	250	CHMI
	Dyje	Znojmo	14/08/2002	4.64	379	150	CHMI
	Jihlava	Dvorce	14/08/2002	2.36	58	50–100	CHMI
Germany	Elbe	Dresden	17/08/2002	9.40		200–500 (1000)	BAFG, (IKSE)
	Elbe	Tangermunde	20/08/2002	7.67		200–500	IKSE
	Mulde	Dessau	15/08/2002	6.25		1000	BAFG, IKSE, SLUG
	Danube	Passau	13/08/2002	10.80		50	BAFG
Austria	Enns	Steyr	12/08/2002	4.85	3200		BLFUW
	Salzach	Salzburg	12/08/2002	6.50	2300	100	BLFUW
	Danube	Linz	13/08/2002	7.50		50	BLFUW
	Danube	Krems	15/08/2002		10000	100	BLFUW

BAFG, Germany Federal Institute of Hydrology; BLFUW, Austrian Ministry for Agriculture, Forestry, Environment and Water; CHMI, Czech Hydrometeorological Institute; IKSE, International Commission for the Protection of the Elbe; SLUG, Saxony State Office for Environment and Geology.

(Table 2.1). It is likely that such coarse estimates were given as an initial reaction, so that flows far exceeding existing 1 in 100 year return period flow were labelled as 200, 500 or 1000 year return periods.

The return period values were quoted widely in the media without any explanation of what they actually meant or how they were derived. On German television, many channels simply reported the event as the *Jahrhundert Hochwasser* ('the 100 year flood'). Once flow data had been collected following the event, a more detailed analysis of the return periods was undertaken giving revised estimates for the Vltava at Prague of 240–800 years (Holicky and Sykora 2003), depending on the extreme value method used.

2.2 Extreme value analysis

With so many return period estimates being put forward during and since this flood event, it is of key importance to understand how such estimates are calculated in order to get the best representation of the severity of the event. As stated at the beginning of this chapter, to assign probabilities and hence return periods to events, it is necessary to apply some mathematical assumptions, often by making the mathematical form of the chosen density function (as defined in Chapter 1) explicit. Unfortunately, there are a vast number of different forms of the distribution, and sometimes quite subtle differences in assumptions can lead to very different return period estimates, as we encountered in the aforementioned anecdote and we will see explained in detail later.

A minor revolution in statistics in the 20th century was the discovery that if the probability of the *largest values* of a set of observations that can be considered random is to be quantified, then there are *only three* kinds of distribution that can arise – the Gumbel (Type I), Frechet (Type II) and Weibull (Type III) distributions, the *extreme value distributions*. This is remarkable because it fixes the mathematical form of the distribution for the extremes of the observations, without having to know the explicit form of the distribution of the observations themselves.

Although extreme value theory applies to an infinite number of observations, in practice, the maximum of a large number of observations (e.g. the daily observations of river flow over 1 year) will closely behave as the theory dictates. Therefore, one practical approach is to use *block maxima*, that is the maximum of a hydrological variable spanning a certain time range (block) such as the annual maximum flow and to fit the extreme value distribution to these maxima. How this method can be used to estimate return periods of extreme events is illustrated in the next section.

2.3 Simple methods of return period estimation

In its simplest form, the return period for extreme events can be calculated by estimating the distribution of the extremes of the observations. By ranking (sorting) the extremes in ascending order, a simple estimate of this distribution can be constructed, and the probability at these extreme values is the so-called *Weibull plotting formula*:

$$P_r = \frac{r}{N+1} \tag{2.1}$$

where r is the rank (order) of the rth extreme observation, numbering from 1 (smallest) to N (largest), and N is the number of extremes. The associated return period, R_r, is:

$$R_r = 1/(P_r) \tag{2.2}$$

If instead, we have the probability of *exceedance*, P_e, then (2.2) becomes $R_r = 1/(1 - P_e)$. The return period unit is the duration of the block.

An example of using this approach is shown in Table 2.1 based on annual block maxima for the Vltava at Prague, with (i) and without (ii) the 2002 flood peak flow. The calculated values here show that the method is not particularly robust in that a flow 4500 cumecs is given a 175 year return period and then a flow that is 750 cumecs (approximately 16%) higher is assigned a 176 year return period. Despite the name, formula (2.1) is universal; it does not depend on the form of the extreme value distribution.

As with all mathematics applied to real data, vigilance is required when making return period predictions. In this context, many variations of (2.1) have been proposed, including the *Gringorten formula* (Gringorten 1963):

$$P_r = (r - 0.44)/(N + 0.12) \tag{2.3}$$

This formula is an attempt to extend the applicability of the formula (2.1) for return periods that are close to the duration of the record or for return periods close to the duration of the block; the numbers 0.44 and 0.12 just scale P_r to be closer to 0 than (2.1) when $r = 1$ and make P_r closer to 1 than the Weibull formula when $r = N$. Typically, (2.1) and (2.3) produce very similar estimates of return periods, *except* at very high non-exceedance probabilities, in which case, the return period estimates can differ substantially (Makkonen 2006). Indeed, for the Vltava data in Tables 2.2 and 2.3, the Weibull formula estimates the return period for flows up to rank 166 ($Q_{max} = 2503$ cumecs) as $R_{166} = 17.6$, but (2.3) gives $R_{166} = 18.3$ years. However, for the full 175 years of record ($Q_{max} = 5300$ cumecs), Weibull is $R_{175} = 176.0$ years and Gringorten is $R_{175} = 312.7$ years.

This disagreement in return period at the largest extreme flows when using the two different formulas will have considerable implications for public perception, shaped by the media, of the significance of any flooding caused by the event, the subsequent pricing of insurance premiums, the design of new flood defences and the eventual costs that those affected will have to bear. Yet, the mathematical arguments on which formula is the 'correct' one are subtle and cannot be decided by looking at the data or the predicted return periods. Unless other information is available that can resolve the conflict (e.g. evidence of more extreme observations

Table 2.2 Ten largest return periods for observed flows on the Vltava at Prague, 1827–2002 (including the events of 2002), estimated using the Weibull plotting formula (2.1).

Year	Annual Maximum Daily Flow (cumecs), Q_{max}	Number of Maxima, N	Rank, r	Non-exceedance Probability, P_r	Return Period (years), R_r
2002	5300	175	175	0.0057	176.00
1845	4500	175	174	0.0114	88.00
1890	3975	175	173	0.0170	58.67
1862	3950	175	172	0.0227	44.00
1872	3330	175	171	0.0284	35.20
1940	3245	175	170	0.0341	29.33
1830	2840	175	169	0.0398	25.14
1900	2770	175	168	0.0455	22.00
1876	2674	175	167	0.0511	19.56
1920	2503	175	166	0.0568	17.60

Table 2.3 Ten largest return periods for observed flows on the Vltava at Prague, 1827–2001 (excluding the events of 2002), estimated using the Weibull plotting formula (2.1).

Year	Annual Maximum Daily Flow (cumecs), Q_{max}	Number of Maxima, N	Rank, r	Non-exceedance Probability, P_r	Return Period (years), R_r
1845	4500	174	174	0.0057	175.00
1890	3975	174	173	0.0114	87.50
1862	3950	174	172	0.0171	58.33
1872	3330	174	171	0.0229	43.75
1940	3245	174	170	0.0286	35.00
1830	2840	174	169	0.0343	29.17
1900	2770	174	168	0.0400	25.00
1876	2674	174	167	0.0457	21.88
1920	2503	174	166	0.0514	19.44
1847	2470	174	165	0.0571	17.50

can be obtained from another source), the safest course of action in these situations is not to trust return periods for the most extreme flows calculated by *either* method entirely or else to report *both* return periods with details of how they were calculated, so that subsequent decisions are made taking into account the uncertainty (see chapter 6).

Both formulas are easy to use and simple to convey but one major disadvantage is that the largest return periods are determined entirely by the period of the record.

So for a 30 year record, the largest possible return period would be 1 in 30 years, and there would be no way of deriving maximum flows for a higher return period. The Weibull formula (2.1) is however widely used in probabilistic modelling when a large number of flood scenarios are available (e.g. more than 1000 scenarios are generated representing a very long time period, say 10,000 years).

2.4 Return periods based on distribution fitting

More advanced techniques allow better flexibility in calculating return periods using more sophisticated mathematics that fit a distribution to the observations or their maxima. Past investigations have shown that certain distributions are suited to known hydrological variables, for example, annual rainfall totals are well fitted by the *gamma distribution* (see the glossary). Hydrological services try to ensure that consistent results are obtained for different gauging stations by using a common distribution for particular hydrological variables.

There are a great many statistical approaches to accomplishing this fitting problem; widely used examples include *maximum likelihood*, the *method of moments* (MOM) and *regression*. The first approach has widespread use in the statistical community; a modified example of the MOMs is described in the Flood Estimation Handbook (Institute of Hydrology 1999) for fitting a distribution to flood observations and least-squares regression is commonly used in hydrological practice. As the latter approach is ubiquitous and intuitive, it is described here as applied to the Gumbel (Type I) extreme value distribution which, for reasons explored in extreme value theory (beyond the scope of this book), is most often used in practice.

The Gumbel extreme value distribution has the following distribution function (the exceedance probability function):

$$P(x; \mu, \sigma) = \exp\left(-\exp\left(-\frac{x-\mu}{\sigma}\right)\right) \tag{2.4}$$

where x is the observed hydrological variable, μ is the *location* parameter of the distribution and σ is the *scale* parameter of the distribution; the notation $P(x; \mu, \sigma)$ simply indicates that x is a random variable and μ and σ are parameters of the distribution; the semicolon separates the variable from the parameters. Fitting this *two parameter* distribution requires finding the optimum values of μ and σ given the observed annual maxima of the hydrological variable.

Regression approaches to fitting extreme value distributions are based on applying some mathematical transformation to the Weibull plotting position (2.1) of the observed maxima so that the cumulative distribution function is a straight line when plotted against the maxima. In the Gumbel distribution case, the transformation 'undoes' the double exponentiation by applying the following in order: log, minus, log, minus to both sides of (2.4). This leads to the following equation:

$$y = -\ln[-\ln(P(x;\mu,\sigma))] = \frac{x-\mu}{\sigma} \qquad (2.5)$$

and both sides equate to a new quantity, which we can label y, which is known as the *reduced variate*. With a little further algebra, this reduced variate can be expressed in terms of x:

$$x = \sigma y + \mu \qquad (2.6)$$

Bear in mind that the term y in (2.6) is identical to the left-hand side of (2.5). The critical point about (2.6) is that it can be recognized as a straight line where y is considered as the independent variable. So, by applying this transformation to the probability (Weibull plotting position) of each maximum, we get an associated value of the reduced variate y, and these points all lie on a straight line when plotted against the observed maxima. This makes it easy to extend a straight line through these points, and so use the slope and intercept to determine the Gumbel scale and location parameter σ and μ.

Given, these facts, we can propose a sequence of operations for Gumbel extreme value analysis:

1 The annual maximum flows are ranked from lowest to highest, labelling the rank r so that the lowest flow has rank $r = 1$ and the rank is increasing with increasing flow.
2 The plotting position probability value P_r is computed for each rank as $P_r = r/(N+1)$ (2.1) where N is the total number of flow values.
3 The reduced variate values y are calculated as $y = -\ln(-\ln P_r)$.
4 The reduced variate values, y on the y-axis, are plotted against flows x on the x-axis, with the line of best fit included (see Figure 2.2).
5 From the line of best fit $y = mx + c$, with slope m and intercept c, we can now calculate the Gumbel parameters $\sigma = 1/m$ and $\mu = -c/m$.
6 Now, we can insert the flow values x back into the Gumbel distribution formula (2.4) to get the associated exceedance probabilities P_e predicted by the Gumbel model.
7 Finally, we can use these modelled exceedance probabilities to obtain estimated return periods using the equation $R_r = 1/(1-P_e)$.

Note that the equation in step 7 is different from Equation (2.2) which relates return period to non-exceedance probability. The probability derived from step 6 is the exceedance probability; therefore, this has to first be subtracted from 1 to derive the return period. Also, as always with such sequences of calculations it is always good practice for the magnitudes of specific calculated return periods to be compared with the input data. If for example the 1 in 10 year flow was calculated at 1000 cumecs where the input data over a period of 50 years had a

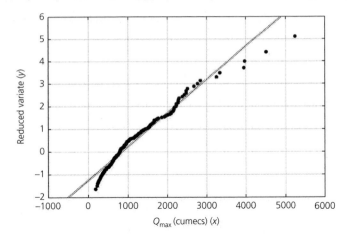

Figure 2.2 Gumbel plot of the annual maximum daily flows on the Vltava at Prague, 1827–2006. The dotted line is the best straight line obtained by least-squares fitting.

Table 2.4 Return periods to two significant figures for observed flows on the Vltava at Prague, estimated by fitting the Gumbel extreme value distribution to the annual maximum daily flow, 1827–2002.

Annual Maximum Daily Flow (cumecs), Q_{max}	Reduced Variate, y	Exceedance Probability, P_e	Return Period (years), R
1000	0.2373	0.45441	1.8
2000	1.7228	0.83647	6.1
3000	3.2083	0.96038	25
4000	4.6938	0.99089	110
5000	6.1794	0.99793	480
6000	7.6649	0.99953	2130

highest observed flow of 200 cumecs, an order of magnitude lower, it is obvious there is something wrong with the calculations and these should be checked.

In Figure 2.2 the Vltava data is used to create such a plot. Here, least-squares regression is used to find the parameters $\mu = 840$, and $\sigma = 673$. Plugging these parameter values into (2.4) gives estimates for the exceedance probabilities and return periods listed in Table 2.4.

This approach obtains another set of estimates for the return periods of the flow. This estimate for $Q_{max} = 5250$ cumecs on the Vltava at Prague is 1 in 700 years, which is much larger than the value reported by the Czech state hydrological service in Table 2.1, and the values are calculated using the Weibull and Gringorten formulas. How reliable is this new estimate? Looking at the plot in Figure 2.2, it is fairly obvious that only the largest maxima actually lie on a straight line.

If we exclude a few of the smaller maxima (e.g. the smallest 20), then the best fit line has a smaller slope and the resulting return period estimates are much closer to those given in Table 2.1. Is this truncation of the data a reasonable adjustment to make? Here the answer is most likely yes, because, as discussed earlier, the theory of extreme values is only a good approximation of the probabilities of the very largest values. On the other hand, removing too many of the smaller values risks not leaving enough observations to get a good straight line fit or systematic error in the recording of very strong flows might come to dominate.

Sometimes, such plots also show the reduced variate on the x-axis (from which the return period is calculated) in addition to the return period. In that case, the slope and intercept of a straight line fit on this plot are simply the Gumbel parameters σ and μ respectively. This approach is common in many software packages such as the UK Flood Estimation Handbook WINFAP software (Institute of Hydrology 1999). The return period is displayed as a secondary x-axis (Figure 2.3).

In the case of uncertainty relating to the data and model applied, there are statistical methods that can quantify the appropriateness of the assumption that the Gumbel distribution applies to this data, which are discussed in Chapter 6, although purely from an assessment by eye of Figure 2.1, it may be concluded that there are better statistical models. Other types of distribution are discussed briefly in the following. The mathematical details of these are beyond the scope of this book, and the reader should just note that they are all methods for estimating a specific return period flood (or indeed other hydrological event such as rainfall)

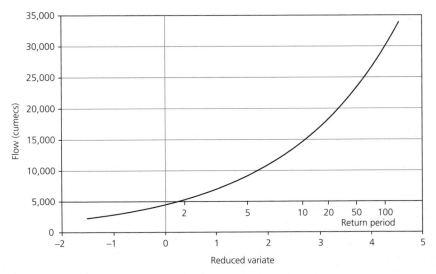

Figure 2.3 An example of the flow (y axis) plotted against reduced variate (x axis) and return period.

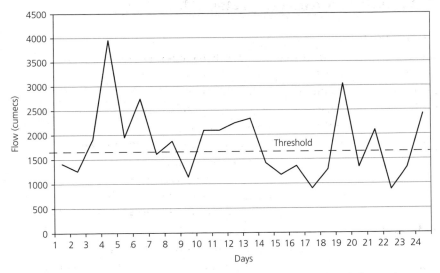

Figure 2.4 An illustration of selecting peaks over a threshold for a period of observations.

from a set of observations over a particular time. Many commercial hydrological modelling software packages, such as the Flood Estimation Handbook (Institute of Hydrology 1999), have automated routines for fitting such distributions.

Other distributions which could be used are the generalized extreme value (GEV) II (Fréchet) and GEV III (Weibull), as described by Hosking et al. (1985) but also other forms which are not extreme value distributions. The *generalised Pareto distribution* (GPD – Pickands 1975; Hosking and Wallis 1987) is used when the values are not block maxima (e.g. the highest values for each year over a period of many years) but are extracted as peaks over a threshold (POT) as shown in Figure 2.4. In the United Kingdom, the *generalised logistic distribution* (GL – Balakrishnan and Leung 1988) is often used in hydrological practice as in certain situations, and GEV and GPD models are found to be less accurate when fitting extreme values than the GL distribution. The GL distribution has a particularly simple mathematical form and is similar in shape to the normal distribution (i.e. the bell shaped curve where observed values are evenly distributed around the mean), but it attributes more probability to extreme positive and negative values than the normal distribution and can allow for features such as asymmetry in positive versus negative values. The *log-normal* (LN – Limpert et al. 2001) distribution is also used in some circumstances, in particular where the values of a variable become very large so by taking the log the relative difference in values becomes smaller and more manageable. The *Log Pearson III* distribution (Cohn et al. 1997) is commonly used in North America, Australia and China; it has three parameters to describe the reduced variate, as opposed to the two-parameter formula shown in Equation (2.5). Note that for these models, the reduced variate Equation (2.5) will take on a different form for each specific distribution.

2.5 Techniques for parameter estimation

The values for the parameters associated with the distributions discussed in the previous section, such as the location and scale parameters shown in Equation (2.5), need to be optimised to ensure the best fit for the distribution (as shown in Figure 2.2). Some standard mathematical approaches are used for this but they often have terminologies which confuse non-mathematicians. One such technique is the use of *statistical moments* (as introduced in Chapter 1). These are simply summary quantities that can be used to describe the distribution of a random variable. Example moments are the mean (average), variance, skewness and kurtosis. The average is commonly understood as the 'central' value of a distribution, the variance is often a meaningful measure of the 'width' of the distribution about an average, the skewness is a measure of the asymmetry of the distribution and the kurtosis the 'sharpness' at the peak of the distribution. These quantities often depend upon one or more parameters of the distribution. In some special cases, there is a unique mathematical relationship between each parameter and a corresponding moment (this is true of the normal distribution, for example). When this occurs, it is possible to fit the distribution to data using the moments to calculate the parameters. This is known as the Method of Moments MOM.

Alternative summaries of the data are derived from *order statistics* introduced at the end of Chapter 1. These include the median, quartiles and the maximum or minimum. These quantities can be estimated from the data by sorting and selection. It turns out that these summary quantities are quite robust to contaminated data, for example, whereas the mean can be significantly distorted by a large error in only one data point, the median is not distorted so easily and therefore may be a more meaningful measure of the 'centre' of a probability density. Fitting the distribution to the data by matching order statistics forms the basis of a technique similar to the MOMs given the name *L-moments* (to denote *linear combinations* of order statistics).

2.6 Bayesian parameter estimation

Another commonly used parameter estimation technique makes extensive use of *Bayes* rule, otherwise widely known as *Bayesian statistics*. Bayes rule is a theorem in basic probability: if the probability of one random variable, X, which depends upon another (in the mathematical jargon, *conditioned upon*), Y, is known, then the probability of Y conditioned on X can also be found, if the probability of X and Y alone, and their joint probabilities (that is the probability of both X and Y co-occurring), are known. Bayes rule is the probability 'calculus' by which this conditioning can be reversed. For the conditional probability, we use the notation $P(X|Y)$ to denote the probability of X conditioned on Y. Then, Bayes rule can be written as:

$$P(Y|X) = \frac{P(X|Y)P(Y)}{P(X)} \qquad (2.7)$$

Perhaps the most essential distinguishing feature between Bayesian and the other, non-Bayesian approaches discussed in this chapter, is that in non-Bayesian approaches, the model parameters to be estimated (such as the Gumbel σ and μ parameters) are not assumed to be random quantities; in principle, they take on a fixed value, and we typically seek the best estimate of that value given the data for X. By contrast, in the Bayesian approach, these parameters are assumed to be random variables. This means that, along with the data X, they must also have a certain distribution. In that case, we might write $P(x|\mu, \sigma)$ instead of $P(x; \mu, \sigma)$ in (2.4) to indicate that the distribution of the data depends upon that of the parameters μ and σ.

If we consider the simple case when we have only a single parameter distribution $P(x|\mu)$, then a Bayesian approach would also propose a distribution $P(\mu)$. In the Bayesian terminology, $P(x|\mu)$ is the 'likelihood' of the data given the parameter value μ, and $P(\mu)$ (the 'prior') is the distribution of the parameter we would expect *before* (*prior to*) seeing the data for X. By using Equation (2.7), we can then attempt to find $P(\mu|x)$, which is known as the 'posterior' distribution of the parameter given the data (note that, in principle at least, we can always find this posterior distribution if we just know the prior and likelihood distributions, by 'integrating out' the distribution over the data $P(x)$, although in practice, this integration may be very difficult to perform).

The Bayesian approach makes intuitive sense if we appreciate that the posterior distribution, that is the distribution of the parameters given the data, is what we actually care about in practice, not the distribution of the data given fixed values of the parameters (the likelihood). There are many other reasons to prefer this approach, particularly when there is only a small amount of data. In that case, the prior distribution naturally acts to constrain the most probable values of the parameters to a meaningful range, for example. Another way of explaining the Bayesian approach is that it makes the estimated parameter values less sensitive to large fluctuations in the data which may be spurious. It also provides a distribution over the parameters, from which we can estimate not only the optimal value given the data but also a *confidence interval* for this estimate. Nonetheless, the Bayesian approach is considerably more complex than non-Bayesian approaches, and there are some thorny questions to address, such as how to choose the prior distribution, which introduces an additional element of subjectivity which could have a large impact on the posterior distribution.

2.7 Resampling methods: bootstrapping

Earlier, we described how to estimate a parameter value given some data, for example the Gumbel parameters μ, σ. Any one draw of a set of data is just a single

Table 2.5 Example bootstrap computations to estimate standard deviation of the average depth of rainfall from a synthetic time series.

Data point index i	Original Data x_i	Bootstrap Data 1	Bootstrap Data 2	Bootstrap Data 3	Bootstrap Data 4	Bootstrap Data 5
1	0.68	0.47	0.74	0.20	0.19	0.53
2	0.74	0.68	2.50	2.50	0.19	0.19
3	0.20	0.53	0.53	0.74	0.19	2.50
4	0.08	0.53	0.74	0.20	0.47	0.08
5	0.53	0.53	0.08	0.68	0.20	0.19
6	0.12	2.50	0.19	0.12	2.50	0.53
7	0.19	0.08	0.74	0.19	0.20	0.25
8	2.50	2.50	2.50	0.12	0.74	0.25
9	0.25	0.53	0.20	0.53	0.19	0.20
10	0.47	0.68	0.47	0.19	0.53	0.19
Average $\bar{x}$	0.58	0.90	0.87	0.55	0.54	0.49
Standard deviation of averages	0.18					

Data values are in centimetre. Note that sampling with replacement means that data values from the original set usually appear more than once in each data set.

set of draws from a random variable, and every set of data will be effectively unique. So, having estimated a parameter value, a question naturally arises: how much might these parameter values vary if we had more data? This is obviously a difficult problem but a rather elegant and simple solution is provided by *bootstrapping*: by *sampling with replacement* from the given data. We remove a value from the data uniformly at random, and 'replace' this back into the data set, and repeat this to create a new set of data of any length we choose. Sampling 'with replacement' in this way ensures that each new data set is potentially different from all the rest, and it also means that the same values from the original data will, most likely, appear more than once in each new data set.

These new, generated sets of data can each be used to provide estimates for the parameters. Then, we can summarize the spread of the distribution of these parameter estimates to get an idea about how much this parameter might vary for unseen data – see Table 2.5.

This bootstrapping procedure is deceptively simple but of enormous generality. Nearly everywhere we have a data set, whether synthetic or natural, we can use this procedure to generate new data sets of any length which (approximately) share the same statistical properties as the original data. So, how does this procedure work? The details are somewhat complex, but essentially, we are using the data to estimate a 'bare bones' probability density for the distribution of the data, and then sampling by replacement just draws new samples from that distribution.

Resampling and bootstrapping are essential ingredients in dealing with uncertainty in process-based hydrology, which is explored in Chapters 3 and 6.

References

Balakrishnan, N. & Leung, M.Y. (1988) Order statistics from the type I generalized logistic distribution. *Communications in Statistics – Simulation and Computation*, **17** (1), 25–50.

Cohn, T.A., Lane, W.L. & Baier, W.G. (1997) An algorithm for computing moments-based flood quantile estimates when historical information is available. *Water Resources Research*, **33** (9), 2089–2096.

Gringorten, I. (1963) A Plotting Rule for Extreme Probability Paper. *Journal of Geophysical Research*, **68** (3), 813–814.

Holicky, M. and Sykora, M. (2003). Statistical evaluation of discharges on the Vltava River, Prague. *Proceedings of the International Flooding Conference: Flooding 2002*, Edinburgh, October 2003, pp. 25–28.

Hosking, J.R.M. & Wallis, J.R. (1987) Parameter and quantile estimation for the generalized Pareto distribution. *Technometrics*, **29** (3), 345–355.

Hosking, J.R.M., Wallis, J.R. & Wood, E.F. (1985) Estimation of the generalized extreme-value distribution by the method of probability-weighted moments. *Technometrics*, **27** (3), 251–261.

Institute of Hydrology (1999). *The Flood Estimation Handbook*, **5** Volumes. Wallingford, Oxfordshire.

Limpert, E., Stahel, W.A. & Abbt, M. (2001) Log-normal distributions across the sciences: keys and clues. *BioScience*, **51** (5), 341–352.

Makkonen, L. (2006) Plotting positions in extreme value analysis. *Journal of Applied Meteorology and Climatology*, **45**, 344–360.

Pickands, J. (1975) Statistical inference using extreme order statistics. *The Annals of Statistics*, **3** (1), 119–131.

CHAPTER 3
Mathematics of hydrological processes

3.1 Introduction

This chapter provides examples of a range of hydrological processes which are represented by equations of differing degrees of complexity, from simple arithmetic equations to the use of partial differential equations and integration. The aim is to allow readers to feel comfortable with such equations when they are presented in other textbooks, handbooks, reports or scientific papers. It is often now the case that papers in hydrological journals make fairly heavy use of such equations. One of the references in this chapter uses sample equations from a paper in a hydrological journal of approximately 15 pages containing over 50 numbered equations!

3.2 Algebraic and difference equation methods

Chapter 1 provided the background for the structure of simple equations where the value to be computed can be isolated using the operations of addition, subtraction, multiplication and division, that is making use only of the elementary *algebraic operations*. When some equations are printed in textbooks or scientific papers, the actual mathematics can sometimes be quite basic, for example *mass balance* equations, but a reader with little recent mathematical experience can be put off from gaining a complete understanding of the model because the equations use Greek letters and subscripts which cloud the simple underlying form of the equations used. For example, a simple soil water balance can be in the form:

$$D = P - \Delta S - ET \tag{3.1}$$

where D is the depth of drainage water issuing from the soil, P is the precipitation, ΔS is the change in soil water storage and the symbol ET represents evapotranspiration. Note that, unless the separate variables E and T have been defined and used elsewhere (an example of which we will see later), we would

Understanding Mathematical and Statistical Techniques in Hydrology: An Examples-Based Approach, First Edition. Harvey J. E. Rodda and Max A. Little.
© 2015 Harvey J. E. Rodda and Max A. Little. Published 2015 by John Wiley & Sons, Ltd.

expect ET to refer to a unique variable rather than to denote the multiplication of E by T.

Further terms may be included to give more detail to the model (3.1), although it still remains a mass balance with quantities being either added together or subtracted, for example, irrigation I can be included:

$$D = P + I - \Delta S - \text{ET} \tag{3.2}$$

Also, some terms can be split into component parts, for example the drainage can be divided into surface runoff D_s and percolation to groundwater D_p, and evapotranspiration can be represented as separate evaporation and transpiration terms (E and T):

$$D_s + D_p = P + I - \Delta S - (E + T) \tag{3.3}$$

In some cases, an individual term in the equation above can be represented as a composite expression if, for example, the irrigation is not given as a depth but instead as a volume applied over a certain area, where a volume V, given in m³, is divided by an area A, in m², giving a depth (m):

$$D_s + D_p = P + \left(\frac{V}{A}\right) - \Delta S - (E + T) \tag{3.4}$$

Further complications arise where the mass balance introduces time because the unknown quantities involved become *functions* of time. For example, an equation for snow water equivalent in the snow pack at time point T (which might be presumed to take on positive integer values) is given by Bergström (1976) as:

$$S_N(T + 1) = S_N(T) + P_S(T) + P_R(T) - Q_N(T) \tag{3.5}$$

where $S_N(T + 1)$ is the snow pack depth at the next time point, $S_N(T)$ is the current snow pack depth, $P_S(T)$ is the precipitation as snow of the current time step, $P_R(T)$ is the precipitation as rain which can freeze and contribute to the snow pack and $Q_N(T)$ is the melting of the snow pack over the current time step.

It is important to appreciate that (3.5) is still a mass balance equation, but it relates values of *functions* at different time points rather than simple variables. Therefore, (3.5) actually represents, potentially, an *infinite number* of algebraic, non-functional equations, for example $S_N(2) = S_N(1) + P_S(1) + P_R(1) - Q_N(1)$, $S_N(3) = S_N(2) + P_S(2) + P_R(2) - Q_N(2)$, etc. Such equations like (3.5) are known as *difference equations* or *recurrence relations*, and they have a lot in common with differential equations introduced in Chapter 1, because solving them requires finding a function of the time point T which holds for *all possible* values of T. Such equations are not generally as easy to solve as purely algebraic mass balance equations like (3.1–3.4).

A combination of terms for different time points and terms which are themselves expressions can produce an equation which appears fairly complex, but is nevertheless still just another kind of mass balance difference equation. For example, the UNESCO report on hydraulic structures in the Danube basin (UNESCO 2004) uses the following equation to compute the natural inflow function Q_a to a reservoir:

$$Q_a(T) = Q_e(T) + Q_e(T-1) + \frac{2(W(T) - W(T-1))}{\Delta T} - Q_a(T-1) \qquad (3.6)$$

Here, Q_e is the outflow function and the function W is the reservoir volume, functions of the time point T. But to clarify, the overall form of this equation is just the addition of three terms and the subtraction of one term.

3.3 Methods involving exponentiation

Chapter 1 introduced exponentiation, that is raising a value to a power or self-multiplication. Terms raised to the power are common in equations used in hydrology. For example, the Manning's equation used to calculate the flow in a river channel of given size and slope requires the raising of two of its terms to a power:

$$Q = \frac{AR^{2/3} S^{1/2}}{n} \qquad (3.7)$$

where Q is the flow (cumecs), A is the cross-sectional area (m^2), R is the hydraulic radius (the ratio of cross sectional area to wetted perimeter in m), S is the channel slope (m/m) and n is the Manning's roughness coefficient (no units). Since this equations mixes multiplication with exponentiation, the order of each of the operations is critical (as indeed, it is critical in Equation (3.4) which mixes addition and multiplication). The values should be raised to the power *before* multiplication with the preceding value; to be clear R is raised to 2/3, S is raised to 1/2 and the resulting values are multiplied together and with the value for A. Finally, the result of the multiplication should be divided by n.

For example, using (3.7) correctly for a rectangular stream channel 3 m wide with 1 m high banks with a channel slope of 0.05 m/m and Manning's n of 0.03 gives a flow of 15.9 cumecs. If the area is multiplied by the hydraulic radius and then the product of these two is raised to the power 2/3, Q comes out as only 10.9 cumecs. This value is considerably less than the correct answer and could have significant implications if the flow was to be used for design purposes.

3.4 Rearranging model equations

It is often the case in hydrological textbooks, papers and reports that equations are written in different forms. This is usually when expressions for some of the parameters are replaced with others where the values or data are more easily obtained, or when equations are rearranged to solve for some variables in terms of others. As discussed in Chapter 1 in detail, although showing each step in the transformation of an equation is critical to avoid mistakes, it is not of particular interest to a hydrologist who would ultimately just want to use a formula to calculate a value for a particular problem. Such sequences of transformations in textbooks can be tedious and confusing to those who are not so mathematically inclined, particularly when the actual transformation in each step is not explained and single words such as 'thus', 'hence', and 'so' are used suggestively. This does not make easy reading for those more familiar with written descriptions, but such transformations must be understood step-by-step if the correctness of the resulting formula is to be ensured.

Continuing on the soil–water balance theme, the Penman–Monteith formula is widely used to calculate evapotranspiration. The equation involves the quantity E_a which represents the energy required for evapotranspiration as shown in (3.7). Although this is more a facet of meteorology and thermodynamics, many hydrology textbooks detail rearrangement of this equation through a series of steps which are largely unexplained. The following form is taken from Shaw (1983):

$$E_0 = H - \frac{\gamma E_0}{\Delta} - \frac{\gamma E_a}{\Delta} \qquad (3.8)$$

In this case, the goal is to rewrite the equation such that E_0 is alone on the left-hand side. The problem requires a few algebraic steps because E_0 appears on both the left- and right-hand sides of the equation. As with all mathematical problems, this can be broken down step-by-step, taking care not to introduce errors. The same operation is done to both sides of the equation at each step, keeping in mind the eventual form of the equation that has to be reached. There are an infinite number of ways of approaching this problem, but to start, the Δ on the bottom of the terms in (3.8) can be removed from the right-hand side by multiplying all terms (on both sides, as always!) by Δ. This leads to:

$$\Delta E_0 = \Delta H - \gamma E_0 - \gamma E_a \qquad (3.9)$$

Now, remembering that the goal is to have E_0 alone (on either the left- or right-hand side, we will aim for the left in this case), all those terms containing that symbol on the right have to be moved over to the left. This is done by *adding* the expression γE_0 to both sides, which cancels out the term $-\gamma E_0$ on the right-hand side:

$$\Delta E_0 + \gamma E_0 = \Delta H - \gamma E_a \qquad (3.10)$$

Next, those terms that contain E_0 can be combined into a single term by *factorising* these terms, that is by finding the factors which they have in common (which of course is just E_0), giving:

$$E_0(\Delta + \gamma) = \Delta H - \gamma E_a \qquad (3.11)$$

Finally, our goal is achieved by dividing through both sides by the expression $\Delta + \gamma$ to ensure E_0 is all that appears on the left-hand side:

$$E_0 = \frac{\Delta H - \gamma E_a}{\Delta + \gamma} \tag{3.12}$$

This suffices to allow the calculation of E_0 given the other variables, but this can also be rearranged to write it in terms of the ratio Δ/γ. If the top and bottom of both sides are divided by γ, then the left-hand side is not changed, but the form of the right hand-side is changed:

$$E_0 = \frac{(\Delta/\gamma)H - E_a}{(\Delta/\gamma) + 1} \tag{3.13}$$

Note that the brackets here are critical to ensure that the proper sequence of algebraic operations is used to calculate E_0.

3.5 Equations with iterated summations and products

So far in this chapter, we have encountered only the basic mathematical operators of addition, subtraction, division, multiplication and exponentiation. What can be much more confusing to students and practitioners is advanced mathematical notation using symbols occurring in, for example, iterated summation and products (see Chapter 1, calculus). However, all mathematical notation is, in principle, rigorously defined in terms of simple processes which when explained in words are actually quite easy to understand. Performing the repeated sum of a number of values often looks daunting in mathematical notation but it is simply the case of adding a sequence of numbers from the starting value in the sequence to the end value of the sequence. The sum is denoted by the operator Σ ('S' in Greek, short for 'sum') and has the counting variable and range of values that variable takes on as superscript and subscript, for example:

$$Q_m = \sum_{i=1}^{31} Q_i \tag{3.14}$$

This equation tells us that the total monthly flow Q_m is the sum of the daily flows for that month, that is flows Q_i, where i is the day number, running from 1 to 31 inclusive. The (sample) mean monthly flow is then obtained simply by dividing this value by the number of days:

$$\bar{Q}_m = \left(\frac{1}{N}\right) \sum_{i=1}^{N} Q_i \tag{3.15}$$

Note in this equation, the upper range (the value 31) in (3.14) has been replaced by the variable N, as the number of days per month varies and the mean monthly flow is represented by $\bar{Q}_m$, pronounced 'q em bar'.

This iterated summation notation is widely used to describe hydrological parameters combined over a time period or an area. Other examples include calculating water quality loads from concentration and flow measurements. The load is the mass of a particular contaminant, such as sediment or dissolved chemicals, which is transported with the flow of a river to the catchment outlet. One of the problems is that whereas the flow can be measured continuously through logging the water level and using flow-level rating equations (see Chapter 4) or other means such as ultrasonic gauging, measuring the concentration of the contaminant in a continuous manner is not possible without complex monitoring equipment. Instead, spot samples are taken at regular intervals and the concentrations are measured back in the laboratory. The product of the concentration (mass per unit volume) and the flow (volume per unit time) will give a load for a specific time. Various equations have been used to derive a load for a monitoring period (e.g. 1 year) based on the concentration and flow measurements and taking the sum of these measurements over time, for example:

$$\text{LOAD} = \frac{K}{N} \sum_{i=1}^{N} (C_i Q_i) \tag{3.16}$$

where C_i is the instantaneous concentration for each sample, Q_i is the instantaneous flow at the time of the concentration samples, N is the number of samples and K is a time conversion factor. This summation to compute the load is shown in Table 3.1 using hypothetical data. Assuming one sample per month, the annual load can be derived by multiplying the time conversion factor of 31,536,000 for the number of seconds in a year, giving a value of 3.4×10^9 g/year.

Table 3.1 Load calculation using Equation (3.16) based on concentration and flow data.

Sample Number i	Instantaneous Concentration C_i (g/m³)	Instantaneous Flow Q_i (cumecs)	Instantaneous Load $C_i Q_i$ (g/s)	Average Instantaneous Load $C_i Q_i / N$, with $N = 12$ (g/s)
1	34.2	4.5	153.90	12.83
2	30.2	4.0	120.80	10.07
3	26.8	4.2	112.56	9.38
4	23.6	3.7	87.32	7.28
5	21.5	3.0	64.50	5.38
6	20.7	2.5	51.75	4.31
7	25.9	1.8	46.62	3.89
8	22.6	1.5	33.90	2.83
9	41.6	2.8	116.48	9.71
10	54.8	3.9	213.72	17.81
11	39.7	4.4	174.68	14.56
12	33.5	4.3	144.05	12.00
Sum				110.02

Another example of using the sum notation to derive hydrological parameters is when a catchment average is required from mapped data covering different proportions of the catchment. In the *Flood Estimation Handbook* (Institute of Hydrology 1999), a key parameter used for the estimation of design floods is the standard percentage runoff (SPRHOST – Boorman et al. 1995). This is calculated for a catchment from the different soil types each with its own characteristic percentage runoff (SPR) and covering a portion, *a*, of the catchment area AREA. The formula used is:

$$\text{SPRHOST} = \frac{1}{\text{AREA}} \sum_{n=1}^{29} (a_n \text{SPR}_n) \tag{3.17}$$

In this example, the sum is multiplied by 1/AREA but could also be divided by the AREA which would be equivalent, and the total area is calculated using $\text{AREA} = \sum_{n=1}^{29} a_n$. Equation (3.17) simply represents a *weighted average*; in this case, the average of all the component SPR values weighted by their corresponding catchment area. A visual example is shown in Figure 3.1 and the accompanying data shown in Table 3.2.

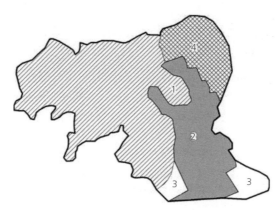

Figure 3.1 SPR values for different soil types 1–4 within a hypothetical catchment.

Table 3.2 SPRHOST values in Equation (3.17) for different soils for the hypothetical catchment shown in Figure 3.1.

Soil Type	Area a_n (km^2)	Standard Percentage Runoff (SPR), SPR$_n$ (%)	Catchment Area-weighted SPR, a_nSPR$_n$
1	3.50	33.8	118.3
2	1.50	25.3	38.0
3	0.12	14.5	1.7
3	0.24	14.5	3.5
4	0.81	44.3	35.9
Sum	6.17		197.4

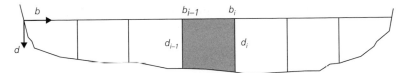

Figure 3.2 Calculation of channel flow by combining subdivision measurements of width, depth and velocity.

Using Equation (3.17), the SPRHOST for the hypothetical catchment with a total area of $6.17\,\text{km}^2$ is calculated as 32.0%.

The iterated summation notation often occurs in producing an observed flow measurement when velocity and depth measurements are made at a number of sub-divisions across a channel cross-section (Figure 3.2).

In a method described by Shaw (1983), the average velocity over successive spatial measurements (v_i, v_{i-1}) and depths (d_i, d_{i-1}) taken from each subdivision boundary is multiplied by the width of the subdivision $b_i - b_{i-1}$, giving the product of velocity and area a_i or the flow q_i for this subdivision. The flow Q, for the whole channel with N subdivisions, is then the sum over all subdivisions:

$$Q = \sum_{i=2}^{N} q_i = \sum_{i=2}^{N} v_i a_i = \sum_{i=2}^{n} \frac{(v_{i-1} + v_i)}{2} \frac{(d_{i-1} + d_i)}{2} (b_{i-1} - b_i) \tag{3.18}$$

As with the sum of a series of values, the iterated *product* of several values is represented using the big Π operator (the Greek 'p' short for 'product'):

$$\prod_{i=1}^{N} x_i = x_1 \times x_2 \times x_3 \times \cdots \times x_N \tag{3.19}$$

This notation is not so common in hydrological studies or models. One example of the use of this notation is in the *Flood Estimation Handbook,* Volume 5 (Bayliss 1999), where the FARL (Flood Attenuation from Reservoirs and Lakes) index is derived for a basin by taking a product of all of the α parameter values which describe the attenuation for each single lake or reservoir based on the surface area of the water:

$$FARL = \prod_{i=1}^{N} \alpha_i \tag{3.20}$$

3.6 Methods involving differential equations

A further level of mathematical sophistication, and a corresponding increase in complexity, occurs when dealing with differential equations. As with difference equations introduced earlier, these are used to describe a dynamic process which is commonly the case in all facets of hydrology as water is continually flowing or moving or changing forms throughout the entire hydrological cycle. The notation, the underlying concepts of the derivative (which expresses the instantaneous rate of change of one variable with respect to another variable, in this case time) and the use of the derivative and partial derivative to construct differential equations have been explained in Chapter 1.

Differential equations are used to model fundamental hydrological processes so they will be encountered in textbooks, scientific papers and software user manuals. However, it can be very difficult to fully grasp these processes without a proper understanding of what the differential equations represent. Unfortunately, solving differential equations requires mathematical sophistication which is often confined to an education in mathematical physics or applied mathematics. The aim of this section is not to train readers to be able to solve any differential equation – indeed, most cannot be solved using simple algebraic manipulations – instead, it aims to provide an alternative explanation so that, at least, the reader is equipped to understand what the differential equation is telling them about the behaviour of the underlying process.

Differential equations occur widely in hydrodynamic modelling – that is simulating the flow of water through, for example, a river channel. These can be used to predict the maximum water levels at particular points along a river reach during a design flood event (e.g. 1 in 100 years). Such models provide the essential design criteria for many hydraulic engineering structures such as bridges and flood defences and also form the basis of flood risk mapping exercises. Looking at the user manual or help files associated with these models, the user will be presented with much detail on the theoretical background. For example, the user manual for hydraulic modelling software (Halcrow/HR Wallingford 1997) presents the *shallow water* or *St. Venant* equations which are used to describe the flow of water in open channels:

$$\frac{\partial Q}{\partial x}(x,t) + \frac{\partial A}{\partial t}(x,t) = q(t) \qquad (3.21)$$

This is an example of a *continuity equation*, where x is the longitudinal channel distance, t is the time, Q is the flow as a function of time and distance, A is the cross-sectional area as a function of time and distance and q is the outflow (which is indicated here as function of time but could also be constant and hence independent of time).

The question arises though as to what does this equation actually represent? It is simply another balance equation: the rate of change of volume over distance, plus the rate of change of area over time, equates to the outflow (which may or may not be constant). As a reminder, this is a partial differential equation because Q and A are functions of both time and position and the continuity equation involves both time and spatial rates of change.

Equation (3.21) does not provide sufficient information to solve for the primary variable of interest, that is the flow. The other governing equation in the St. Venant model is known as a *momentum equation* which provides sufficient constraints to solve:

$$\frac{\partial Q}{\partial t}(x,t) + \frac{\partial}{\partial x}\left(\frac{\beta Q^2}{A}\right)(x,t) + gA\frac{\partial H}{\partial x}(x,t) - gA(x,t)S_f(x,t) = 0 \tag{3.22}$$

where in addition to the variables for (3.21), H is the water surface elevation above a datum, β is the momentum correction coefficient, g is the gravitational acceleration and S_f is the friction slope.

One particularly noteworthy aspect of (3.22) is that the right-hand side is zero. For students or practitioners who have often used a calculator to input parameter values to an equation and press the "=" button to get an answer, having an equation in a form where the "answer" is zero seems nonsensical. But referring back to Chapter 1, this is an implicit equation where not all the terms are known, which by rearranging, could also be written with one set of components equal to another set , namely:

$$\frac{\partial Q}{\partial t}(x,t) + \frac{\partial}{\partial x}\left(\frac{\beta Q^2}{A}\right)(x,t) + gA\frac{\partial H}{\partial x}(x,t) = gA(x,t)S_f(x,t) \tag{3.23}$$

Typically, Equations (3.21) and (3.22) will be solved together numerically by breaking the x- and t-axes up into a set of fine grid points, and replacing the derivatives with their finite difference counterparts, for example $(Q(x+\Delta x,t) - Q(x,t))/\Delta x$ for $\frac{\partial Q}{\partial x}(x,t)$. Then, the equations can be solved as $Q(x+\Delta x,t)$ in terms of $Q(x, t)$ on the grid to get explicit formulas to compute an approximate solution to the differential equations.

3.7 Methods involving integrals

In general, equations involving integrals are not such a regular feature of equations of hydrological processes since most equations are used to describe rates, that is tracking the flow of a mass of water at a particular instant in time. Integration on the other hand combines separate units of mass and other hydrological quantities together, over a given time interval, as described in detail in Chapter 1. Nonetheless, as described in Chapter 1, these differential and integral ways of representing the same physical quantities and laws can always be transformed into each other, but the differential form generally leads to more easily solved equations.

Mansell (2003) used the integral form to describe the method of dilution gauging, where a known quantity of a chemical which can readily dissolve (such as salt) is added to a stream either directly or as a solution with a known concentration. The concentration of the salt is then measured downstream at regular intervals, usually at a high-time resolution (e.g. 1 minute). Assuming that the stream flow (Q) is constant over a short length of channel, the mass flow (M) of the salt (i.e. the flow, Q, multiplied by the concentration, C) is:

$$M(t) = QC(t) \tag{3.24}$$

Integrating both sides of this equation with respect to t, we get:

$$\int M(t)\,dt = Q \int C(t)\,dt \tag{3.25}$$

As the aim is to calculate the flow, using algebra, the above equation can be rearranged to solve for Q:

$$Q = \frac{\int M(t)\,dt}{\int C(t)\,dt} \tag{3.26}$$

Further refinements to the equation can be made because there is normally a background concentration C_0 (not changing with time) which means the actual observed concentration in the river $C_1(t) > C_0$ (as a function of time) is higher than the initial concentration of the solute, C_0:

$$C(t) = C_1(t) - C_0 \tag{3.27}$$

This term can be substituted back into (3.26) and we get:

$$Q = \frac{\int M(t)\,dt}{\int (C_1(t) - C_0)\,dt} \tag{3.28}$$

More often, differential equations have a range of values to which they apply, similar to the sum of equations described in Section 3.3. Integral equations in this form are often used in model reports, handbooks or papers.

References

Bayliss, A. (1999) *Catchment Descriptors. Flood Estimation Handbook,* Vol. **5**. Institute of Hydrology, Wallingford, Oxfordshire.

Bergström, S. (1976) *Development and Application of a Conceptual Runoff Model for Scandinavian Catchments,* Bulletin Series A No. 52. Department of Water Resources Engineering, Lund Institute of Technology, University of Lund, Lund.

Boorman, D.B., Hollis, J.M., & Lilly, A. (1995) *Hydrology of Soil Types: A Hydrologically Based Classification of Soils in the United Kingdom.* Report No. 126, Institute of Hydrology Wallingford, Oxfordshire.

Halcrow/HR Wallingford (1997) *ISIS Flow User Manual.* Sir William Halcrow and Partners, Burderop Park, Swindon.

Institute of Hydrology. (1999) *Flood Estimation Handbook,* **5** Volumes. Wallingford, Oxfordshire.

Mansell, M.G. (2003) *Rural and Urban Hydrology.* Thomas Telford Ltd., London.

Shaw, E.M. (1983) *Hydrology in Practice.* Van Nostrand Reinhold, Wokingham.

UNESCO. (2004) Inventory of the Main Hydraulic Structures in the Danube Basin. The Danube and Its Cathcment – A Hydrological Monograph Follow-up Vol. VIII/1, Budapest, Hungary, 62pp.

CHAPTER 4

Techniques based on data fitting

4.1 Experimental and observed data

As a science, hydrology has always been heavily based on the collection of data from field experiments. This can be in the form of observations of rainfall, river flows, groundwater levels or concentrations of pollutants. The processes which relate these variables are often understood conceptually based on physical reasoning but are difficult to represent as an equation like those described in Chapter 3. Therefore, the more common method to define relationships resulting from these observations is to simply plot the variables as an '$x - y$' scatter plot and see what function will fit the data.

This method, known as *regression* in statistical circles, is a common approach for data analysis in many disciplines. For such a procedure to be meaningful, sufficient data points are required, and we need to choose a function with the appropriate level of flexibility for fitting this particular data. The problem to solve is that all data has measurement error, and if the function is too flexible, for example because it has too many free parameters that we want to find by fitting the data, the function will simply reflect the error in this particular data. If this happens, the procedure is worthless because we cannot rely on it to make predictions. As an illustration of this point, for *any* two '$x - y$' pairs, there is a *unique* straight line that goes *exactly* through these two points; a line has two free parameters – the slope and the intercept – and therefore we need many more than two data points to avoid error in the data from influencing the fitted function so badly that the result is useless in practice.

Typically then, the number of data points must be much larger than the number of free parameters. Exactly how many are required depends upon several factors, principally the magnitude of the error in the data and the nature of the function, but this is beyond the scope of this book. For simple functions and typical data, having at least 10 times the number of free parameters usually suffices. Therefore, for a straight line, we would need about 20 '$x - y$' pairs to get good estimates of the slope and intercept.

When making predictions given new 'x' data, it is important that we use the function appropriately. Usually, this means that it is risky to make predictions about 'y' values if the new 'x' data lies far away from the 'x' data used to fit the

Understanding Mathematical and Statistical Techniques in Hydrology: An Examples-Based Approach, First Edition.
Harvey J. E. Rodda and Max A. Little.
© 2015 Harvey J. E. Rodda and Max A. Little. Published 2015 by John Wiley & Sons, Ltd.

function. This is because *interpolation* is much more reliable than *extrapolation*. If we extrapolate from data, we have no nearby '*y*' values with which to verify that the function is giving us meaningful answers. Generally, additional knowledge or insight is required in order to ensure that predictions based on regression are not just a convenient mathematical fiction. Regression is therefore no substitute for relevant physical knowledge of the problem.

4.2 Rating curves

Probably the most widely used regression in the field of hydrology is the practice of deriving flow-level rating curves to provide flow values at gauging stations based on observations of water levels. The measurement of flow in a river has traditionally been a much more difficult task than measuring the water level, largely because it involves making observations from within the channel rather than at the river bank. Observations of water level can be made with relative ease such as using a float installed on a stilling well connected to a data-logger (Figure 4.1) which can provide a continuous record of the water levels logged at a high temporal resolution (e.g. every 15 minutes). A simpler approach can be just reading the level from a stage board. With the exception of the recent ultrasonic gauging and acoustic Doppler techniques, flows in a river have involved taking measurements at different locations across the channel with a flow meter. Such observations cannot be made continuously so instead flow measurements are

Figure 4.1 An example of water level monitoring.

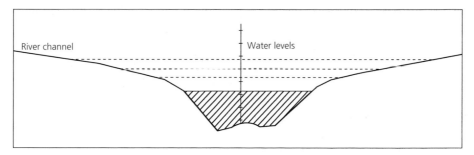

Figure 4.2 The effects of increasing water levels on the area and flow of water in a natural river channel.

made over a number of occasions with different water levels. The data points are plotted as an '$x - y$' scatter plot and a rating curve is derived from this data as the curve of best fit through the points. The equation resulting from the curve fitting is then used to produce a rating table where for each increment of water level, commonly 10 cm, an associated flow is given. The fitting is *nonlinear* (e.g. not simply proportional) as a consequence of the physical assumptions. Also, the channel is not of a regular shape, and as the slope of the banks gets shallower at high flows, a small increase in water level can mean a significant increase in flow (Figure 4.2).

The nonlinear rating equations for flow Q and level H take the following mathematical form:

$$Q = aH^b \tag{4.1}$$

where a and b are constants. On occasions where Q is not zero and when H is zero, due to the position of the stage board or level recording apparatus, a correction factor X is required:

$$Q = a(H + X)^b \tag{4.2}$$

Mathematically, a is known as a *scale factor* or *coefficient* and b is known as an *exponent* (see Chapter 1 for the definition of exponentiation). More than one rating equation can be used when there is a particular channel geomorphology such that water is contained within an inner channel at low flows but then expands to cover the whole channel at higher flows. An example of this is the River Glomma in Norway where at the Nor gauging station two rating equations are used and are as follows:

$$Q = 82.8326(H + 0.63)^{1.374} \quad \text{(for levels } H < 0.75\,\text{m)} \tag{4.3}$$

$$Q = 193.5305(H + 0.03)^{1.6264} \quad \text{(for levels } H \geq 0.75 \text{ and } H < 10.5\,\text{m)} \tag{4.4}$$

The parameters a, b and X would have been obtained from plotting the flows against the levels and finding the curve of best fit. Plots of the rating curves using these equations are shown in Figure 4.3. In this case, the parameters in

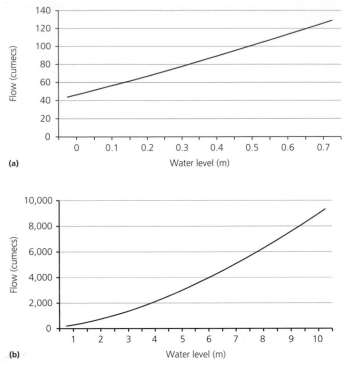

(a)

(b)

Figure 4.3 Rating curves and equations for the Glomma at Nor for (a) levels below 0.75 m and (b) levels between 0.75 and 10.5 m.Data from Norwegian Water Resources and Energy Directorate (NVE).

(4.2) are $a = 82.8326$, $X = 0.63$ and $b = 1.374$, corresponding to (4.3). The degree of precision of these parameters (i.e. up to four decimal places) shown in this example is perhaps too high for the nature of the measurement and is probably a result of the computer software used to fit the curve.

In this particular example, Equations (4.1) and (4.2) are nonlinear: they are not straight lines. This makes fitting quite difficult and the mathematics can become very complex. Standard straight line fitting techniques are far simpler and easier to use. Fortunately, it turns out that (4.1) can be rearranged to make them linear by taking the logarithm of both sides:

$$\ln Q = \ln\left(aH^b\right) \tag{4.5}$$

Using the properties of logarithms, this becomes:

$$\ln Q = \ln a + b \ln H \tag{4.6}$$

So, this tells us that the logarithm of Q is proportional to the logarithm of H – a straight line. Then the slope is b. The intercept (the value of logarithm of Q when the logarithm of H is zero) is the logarithm of a. Therefore, we can use standard straight line fitting techniques to find the best fit values of the parameters a and b.

So far, no discussion of the origin of the particular *form* of the regression function (4.1) has been provided. In this physical situation, it is not unreasonable to start by assuming steady, incompressible flow so that, assuming specific channel geometry, Bernoulli's equation can be manipulated to solve for discharge Q in the channel. The final form for Q is identical to (4.1), but a and b take on specific values arising from the physical assumptions. Experience shows however that with the fixed values of a and b thus obtained, predictions of flow rates Q given measured values of H in actual channels are often unrealistic. This lack of realism can be due to very many complex factors: resolving all of these is usually impossible so one simple response is to introduce some flexibility by allowing a and b to be set by the measured properties of the particular channel. This is by no means the only way to choose the form of the regression function, and in many applications there is no such additional physical insight that might allow us to choose the form of the function. In these cases (and they are numerous), the data itself is all we have to go on in order to choose the form of the regression function.

4.3 Regression with two or more independent variables

A slightly more complex analysis has to be undertaken when trying to relate more than two 'x' variables to a single 'y' value. This is often a problem faced by research hydrologists when trying to derive standard methods to estimate river flows based on two or three factors where values can be easily obtained from observational or mapped data. This type of approach was widely used in the United Kingdom within the Flood Studies Report (FSR; NERC 1975), a comprehensive publication describing the methods used to estimate design floods. Although the FSR has now been replaced by the more up to date *Flood Estimation Handbook* (FEH; Institute of Hydrology 1999), many equations produced from the regressions are still taught in hydrology courses or used by practitioners. One aspect of such equations which makes them stand out when compared with those described in Chapter 3 is the use of numerical constants rather than the purely algebraic representations of formulae such as the Penman–Monteith equation. The selection of numerical values can be particularly puzzling to students or others when viewing equations based on regression – and this sometimes encourages the unquestioning use of 'formulas'" without developing the corresponding understanding of their scope of validity. For example, an equation widely used in the United Kingdom to predict median annual flows (i.e. 1 in 2 year) in small catchments is:

$$Q_2 = 1.08 \left(\frac{\text{AREA}}{100}\right)^{0.89} \text{SAAR}^{1.17}\ \text{SPR}^{2.17} \tag{4.7}$$

where AREA is the area of the catchment in hectares, SAAR is the standard average annual rainfall in mm and SPR is the soil percentage runoff coefficient for five soil classes mapped over the United Kingdom as part of the FSR. The equation has been derived by fitting observed flows to obtain numerical values of the three parameters (AREA, SAAR and SPR) for a number of catchments. Note the parameter SPR here is different from that used in Figure 3.1 and Table 3.2.

As described above for the simpler example (4.1) with one independent variable (*H*), it is generally mathematically easier to fit straight lines (which are *linear*) than general nonlinear curves. In the case of (4.7), the situation is complicated by the existence of three independent '*x*' variables. In the case of one independent variable, it is simple to visualize the relationship between '*x*' and '*y*' variables by plotting them on a graph. This allows us validate the appropriateness of the fitted function against the measured data points. As shown in (4.5), we can make the curve linear by using a *log–log (scatter) plot*, that is, by plotting the logarithm of *H* (or *H* + *X*) against the logarithm of *Q*. An example of a log–log scatter plot (using example data) is shown in Figure 4.4; the linear fit in this case is not implausible.

When there are two or more '*x*' variables, it is not so simple to plot the relationship. In this case, we usually need to rely on purely statistical measures of the 'goodness of fit'. This is explained in more detail in Chapter 6, but we can say that it typically involves calculating the *total error* of the fitted function with reference to the data points.

Fortunately, (4.7) can be made linear using the same 'trick' as for (4.1), by taking the logarithm of both sides which brings the exponent terms into multiplication terms:

$$\ln Q_2 = \ln 1.08 + 0.89 \ln \frac{\text{AREA}}{100} + 1.17 \ln \text{SAAR} + 2.17 \ln \text{SPR} \tag{4.8}$$

This is in the form:

$$y = a + b_1 x_1 + b_2 x_2 + b_3 x_3 \tag{4.9}$$

which shows that this is a purely linear equation in three independent variables. So, the values 0.89, 1.17 and 2.17 appearing in (4.8) are just the gradients of *y* with respect to x_1, x_2, x_3 accordingly.

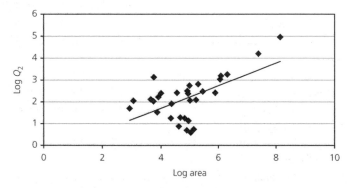

Figure 4.4 A scatter plot of log of Q_2 against log of area for example data (not that used to derive Eq. 4.7).

4.4 Demonstration of decaying quantities

The processes described in Sections 4.2 and 4.3 generally show a positive relationship where one variable increases (e.g. flow) then the other variable also increases (e.g. level). Many other hydrological processes show the reverse that when one variable is increased another variable decreases proportionally. Examples could be the reduction in flow with time after a flood peak has passed (i.e. the recession of the hydrograph) or the reduction in effluent concentration with distance downstream from its source as it becomes more diluted by the ambient water.

Such processes are usually nonlinear in that there is not a constant reduction in the value of one variable as the other increases. It is more common that there is an initial rapid reduction and then a slower decay as the reducing variable returns towards zero or a stable value. An example is presented below from experimental agricultural plots in the United Kingdom (Tyson et al. 1990) showing the decay of inorganic nitrogen in the soil as more is flushed out with the soil drainage (Rodda 1993). In agricultural land, during the growing season (spring and summer) plants will take up water and nutrients such as nitrogen from the soil. At the point of harvest, there is an excess of inorganic nitrogen in the soil from plant residues, which is gradually leached out of the soil in the drainage water throughout the autumn and winter once the soil moisture deficit is replenished.

The rate of decay of the inorganic nitrogen differs for different soil types as shown below. For a well-drained soil (A), there is a more steady decay over time and for a poorly drained clay (B), the majority of the inorganic nitrogen is leached within the first 100 mm of drainage (Figure 4.5).

When equations were derived for the best fit curves from these plots, they took the following form:

$$\text{Soil type (A)}: \quad y = \exp(4.56 - 0.005x) \tag{4.10a}$$

$$\text{Soil type (B)}: \quad y = \exp(4.12 - 0.006x) \tag{4.10b}$$

The exponent function is explained in Chapter 1. The first quantity (i.e. 4.56 and 4.12) determines the initial level when $x = 0$ and the second quantity determines the rate of decay; hence the second plot shows a sharper decay ($0.006 > 0.005$).

Other examples of exponential decay are the decay of nitrate from soil into groundwater before its emergence into a stream developed by Cooper (1990) and expressed in the form:

$$C_s(t) = C_{dp}\, e^{-rt} \tag{4.11}$$

where $C_s(t)$ is the concentration of nitrate in the stream, C_{dp} is the concentration of nitrate entering the groundwater and r is the attenuation coefficient which is a negative power, hence the value of r must be positive otherwise the concentration would grow larger over time.

Equations (4.10) and (4.11) are both in the same form $y(x) = Y_0 \exp(-r \times x)$, remembering from chapter 1 that the exponent can be written as e raised to a

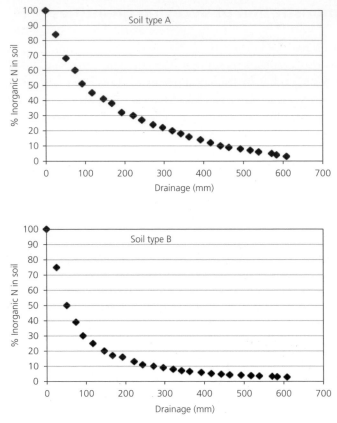

Figure 4.5 The decay of inorganic nitrogen (N) in the soil during autumn and winter drainage for two different soil types.

power or *exp*. The only slight complication is that in Equation (4.10), the initial values $y(0) = Y_0 = \exp(A_0)$ so that $A_0 = \ln Y_0$. All equations such as this can be put into linear form (4.9) by the transformation $\ln y(x) = \ln Y_0 - rx$ which is clearly the equation of a line with slope $-r$ and intercept $\ln Y_0$. In this form, we can use simple straight line fitting to any given set of data.

4.5 Analysis based on harmonic functions

Many hydrological processes have a cyclical form which is often controlled by seasonal factors. When the output of a particular parameter is plotted against time over a year, the parameter can show a growth and then decay such as the plot of soil moisture deficit shown in Figure 4.6 where monthly values over a 3 year period are plotted by month. In the United Kingdom, soils are generally saturated during the winter, and then a soil moisture deficit builds up in the spring as the uptake from plants increases and peaks towards the end of the summer. The moisture is then replenished when plants die in the autumn.

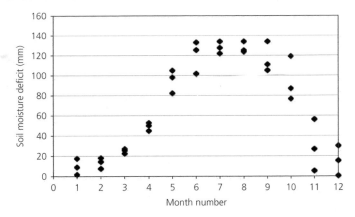

Figure 4.6 Variation of soil moisture deficit over time from January (1) to December (12) for Croydon, near London. Data from the Meteorological Office.

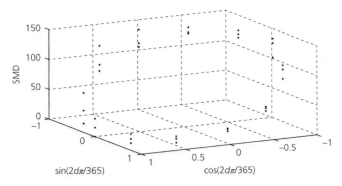

Figure 4.7 3D scatter plot of the data from Figure 4.6 showing how the data lies approximately on a circle embedded in a plane in 3D.

This cyclical behaviour can be demonstrated for many other hydrological variables such as rainfall, stream flow, groundwater recharge, transpiration and nutrient concentrations; some of these may not have such a clearly defined cyclical or seasonal form as soil moisture deficit so a better idea of how the data conforms to a cyclical pattern can be demonstrated by fitting a harmonic function. An example of a first-order harmonic function is as follows:

$$Y(d) = a + b \cos\left(\frac{2\pi d}{365}\right) + c \sin\left(\frac{2\pi d}{365}\right) \qquad (4.12)$$

where Y is the hydrological variable, d is the day number (i.e. 1–365) and a, b and c are constants. This type of equation may appear quite confusing since it includes sine and cosine terms and π. Those not immediately familiar with mathematics will know that sine and cosine are used in trigonometry to calculate angles and lengths of triangle, and π is the ratio of the circumference of a circle to the

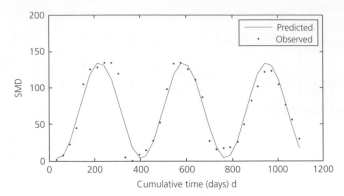

Figure 4.8 Data from Figure 4.6 plotted from the start of the data collection period showing the harmonic fit.

diameter. So, why are they used to fit a line to hydrological variables observed over time? Firstly, the function represents an annual cycle, hence the value of 365 (i.e. the number of days in a year). Representing the year as a full circle, if degrees are measured in radians, one full cycle goes through 2π radians (which corresponds to 360°). Cosine and sine give the magnitudes of the x and y position around the circle, relative to the origin. The circular nature of this harmonic data is evident from the plot in Figure 4.7. The data also lies in a plane due to the linear nature of the function fitting; this occurs because the a, b and c parameters are not inside any other function. If the period of the data was not known (in this case it is known to be 365), then the function fitting would be much more complex and could not be represented as a plane.

If the data are plotted over the full extent of the time period (Figure 4.8), the fit of the harmonic function clearly shows the seasonal variation in soil moisture deficit. Once again, Equation (4.12) is in the linear form $y(x) = a + bx_1 + cx_2$ where $x_1 = \cos\left(\dfrac{2\pi d}{365}\right)$ and $x_2 = \sin\left(\dfrac{2\pi d}{365}\right)$, so that simple plane fitting can be used to estimate the variables a, b and c given some data.

References

Cooper, A.B. (1990) Nitrate depletion in the riparian zone and stream channel of a small headwater catchment. *Hydrobiologia*, **202**, 13–26.

Institute of Hydrology (1999) *The Flood Estimation Handbook*. Wallingford, Oxfordshire.

Natural Environment Research Council (NERC). (1975) The Flood Studies Report, Vol. **5**. HMSO, London.

Rodda, H.J.E. (1993) *The development of a nitrogen cycle model to predict nitrate leaching from grassland catchments in the UK*. PhD thesis, Department of Geography, University of Exeter, UK.

Tyson, K.C., Roberts, D.H., Clement, C.R. & Garwood, E.A. (1990) Comparison of crop yields and soil conditions during 30 years under agricultural tillage or grazed pasture. *Journal of Agriculture Science (Cambridge)*, **115**, 29–40.

CHAPTER 5
Time series data

5.1 Introduction

This chapter has presented some of the different techniques used in hydrology to produce relationships based on data fitting. One of the key hydrological variables however is time, and many hydrological data sets are time based such as the change in flow of a river over time. The time-based data may be at a relatively short time scale such as minutes, hours or days to observe the response of a river catchment to a rainfall event through the event hydrograph or be at a much longer scale such as years, decades or even centuries to consider longer term effects such as land use and climate change. The last part of Chapter 4 discussed some examples of time where a curve could be fitted to define a seasonal distribution of observed values. This chapter considers the nature of time series data, how it is used and analysed in hydrology including the fitting of mathematical functions to identify trends and patterns in the data and the application of mathematical techniques to time series data in an attempt to predict future magnitudes of a given parameter such as flow.

5.2 Characteristics of time series data

Time series data is usually defined as an observation taken at a particular interval over a given range of dates. Figures 5.1, 5.2, 5.3 and 5.4 show examples of such data at different temporal resolution ranging from 1 minute to one decade.

Each figure shows different patterns evident within hydrological times series data; the simple hydrograph rising from a low value to a peak and receding, the spikey nature of rainfall where periods of no rain are interspersed with peaks and general trends over longer periods showing cyclical variations. One of the key features about time series data by comparison to other hydrological data to which models are fitted (as illustrated in Chapter 4) is that the ordering of the observations matters. For example, you could not reverse the data and expect it to make sense, unlike, for example catchment areas, where the order in which these are presented does not matter.

Understanding Mathematical and Statistical Techniques in Hydrology: An Examples-Based Approach, First Edition. Harvey J. E. Rodda and Max A. Little.
© 2015 Harvey J. E. Rodda and Max A. Little. Published 2015 by John Wiley & Sons, Ltd.

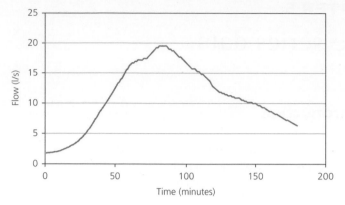

Figure 5.1 Surface runoff measured at 1 minute resolution from experimental 1 ha plots at North Wyke Research, UK (Rodda and Hawkins 2012), for 13 May 2007, from 09:00 to 12:00.

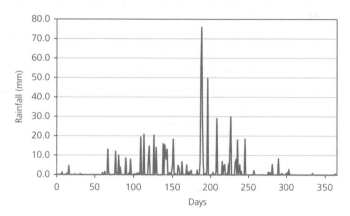

Figure 5.2 Observed daily rainfall during 1998 near Hancheng, China.

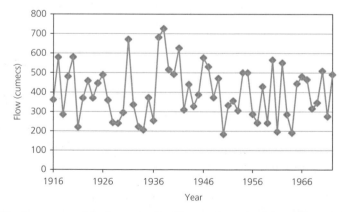

Figure 5.3 Maximum annual discharge on the Morava River, Czech Republic, 1916–1972. From UNESCO (1976).

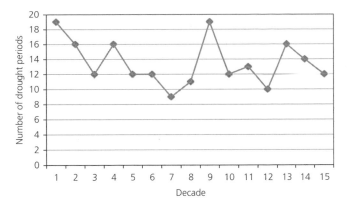

Figure 5.4 Number of meteorological droughts observed at Oxford, UK, per decade from 1853 to 2002.

Some time series such as hourly flow data in a river will be controlled by the history of that data in that the flow will not suddenly jump from a high magnitude to zero within the unit of time but instead would have a correlation with time. So, the current stream flow is actually dependent on previous stream flows, as long as measurements are sufficiently close together. Other time series however can show a lack of correlation in time. Rainfall, for example, when collected over a longer period, such as 24 hours, in certain places has limited or no correlation in time. Looking at a higher resolution such as minute by minute is likely to give a higher correlation in time. These distinguishing features mean that specialized techniques are very often required for the analysis of time series data.

5.3 Testing for time dependence

If a time series has dependence in time, then a data point collected at time t will depend (in the sense of conditional probability at least) on the previous data points $t-1$, $t-2$.... So the probability of some measurement x_t (e.g. mean daily flow in a river) at time t is not independent of previous times, for example $P(x_t|x_{t-1}, x_{t-2}...) \neq P(x_t)$ (the vertical bar indicating conditional probability; see Chapter 1). One simple way of characterizing this dependence is by measuring *correlation in time* known as *autocovariance*. The (unbiased) Pearson autocovariance (R) can be estimated from data using the following formula:

$$R(d) = \frac{1}{N-d} \sum_{i=1}^{N-d} (x_i - \overline{x})(x_{i+d} - \overline{x})$$ (5.1)

where d is the time delay (in samples) and $\overline{x}$ is the estimated mean of the sample data (typically, this will be just $\overline{x} = (1/N)\sum_{t=1}^{N} x_t$) and N is the number of values in the data. To explain this, note that if $d=0$, then $R(0)$ is just the variance (see

Chapter 1). When d is not zero, this formula measures the extent to which the time series, and the series shifted by d time steps, are correlated (vary with each other). In other words, if you were to plot the series x_t against the series x_{t+d} and if the autocovariance $R(d)$ is large and positive, then the plot would be close to a straight, 45° line.

If we normalize this formula by dividing by the variance, we end up with the (sample) *autocorrelation coefficient*:

$$\rho(d) = \frac{R(d)}{R(0)} \tag{5.2}$$

This has the property that $\rho(0) = 1$, and for any other time delay, it lies between -1 and $+1$. This normalization allows us to compare the time dependence of different time series. We can interpret the autocorrelation coefficient as with any correlation: small values (i.e. close to zero, either negative or positive) indicating that there is little, if any dependence in time. Conversely, large positive or negative values indicate that there is considerable time dependence. For example, daily rainfall depth in marked seasonal climates, for example the monsoon, will have large autocorrelation for $d = 1$ and larger, whereas, humid, temperate zones with an even spread of rainfall will have $\rho(d)$ small for $d = 1$ and greater.

Autocorrelation is not the only way to measure serial (time) dependence, but it is the most accessible and easily computed. However, there are many situations in which it is inappropriate, in particular, when the time series has strong outliers, that is occasional values, that deviate very systematically from the rest. In these cases, the autocorrelation can be a severe underestimate of the actual time dependence. Techniques such as the *Spearman rank correlation coefficient* (Wasserman 2003) are less affected by such issues and have formulas that are as simple as (5.1).

5.4 Testing for trends

The simplest approach to detecting continuous trends is when the change is expected to be linear (i.e. the change is increasing or decreasing with a constant amount in each time interval, indicating a linear trend). By plotting the observed data against time and fitting a line using the techniques presented in Chapter 4, a clear relationship between the hydrological variable and the time can be detected. In terms of statistical tests detecting linear trends in the time series, these are often undertaken by correlation testing such as the *Pearson correlation test* (Wasserman 2003). Slightly more complex is the situation where the trend is not linear, but strictly increasing or decreasing; in this case, the *Spearman* or *Kendall tau* correlation test (Wasserman 2003) can be used. More sophisticated tests can check for a general statistical relationship between time and the trend in the data, such as by computing the *mutual information* (Cover and Thomas 2006).

5.5 Frequency analysis

If a time series, say x_t repeatedly comes back to the same value (or very close to that same value) after a time delay d, we say that it is *periodic*. An example would be the tidal depth at the coast, which is tied to a very precise physical oscillation. This would be represented by the equation:

$$x_t \approx x_{t+d} \tag{5.3}$$

for all t. A theoretical example of a time series which has precisely this property is the sine function:

$$x_t = \alpha \sin(\phi t) \tag{5.4}$$

where $\alpha > 0$ is the *amplitude* and ϕ is the *frequency*. Then the period of this time series is 2π (this originates in trigonometry and the angle returning to the same value once per revolution) Of course, no real hydrological time series is this precise. Although it may be periodic, it will not be sinusoidal. However, it turns out that that *any* periodic time series can be represented as a sum of *harmonic* sinusoidal components found using a decomposition known as the *Fourier series*:

$$x_t = \sum_{k=0}^{\infty} (A_k \sin(kt) + B_k \cos(kt)) \tag{5.5}$$

To see how this breaks down, we need to look at the individual term $A_k \sin(kt)$. This is a sinusoid as in (5.4), with amplitude A_k and period k. Similarly, the term $B_k \cos(kt)$ is another sinusoid. *Fourier analysis* is the process by which we take a periodic time series and find the terms A_k and B_k. Every time series has a unique (and infinite) set of these values, which tells us that we can identify each time series by its harmonic amplitude series, A_k and B_k. Of course, given that there are an infinite number, in practice we can only actually calculate a finite number, so the sum in (5.5) would not be infinite. A plot of harmonic number k against amplitudes A_k and B_k is known as an example of a *frequency domain* plot; it contains most of the same information as the time domain plot of t against x_t, but in terms of the amplitude of the sinusoidal components in (5.4). If we find a very large value of a particular A_k, this would indicate that the time series is dominated by a period of k.

Fourier analysis is very closely related to harmonic function fitting as described in Chapter 4. Yet, the application is very different. In harmonic function fitting, we choose the dominant period in advance. Fourier analysis is often used in the situation where we do not know the periodicity of a time series.

Equation (5.5) can be modified for the case of a finite length time series (of course, every practical data set will be finite). In such case, the discrete Fourier transform (DFT), which has a finite summation, is usually applied. We should also point out that the A_k and B_k can be transformed into the so-called *polar representation* with two numbers, phase ϕ_k and amplitude α_k. A plot of α_k against k is therefore known as an amplitude plot, whereas a plot of ϕ_k against k is known as a phase plot. The DFT can be computed very rapidly using an algorithm called the *fast Fourier transform* (FFT); this is the algorithm usually encountered in time series analysis applications. Because the DFT contains all the same information as the original time series, it has a reverse operation known as the inverse DFT (IDFT) which takes a frequency domain representation and converts it back into a time series. There is a corresponding fast inverse transform known as the IFFT.

5.6 Other analysis methods

There are a vast number of different mathematical approaches to analysing time series, not all of which can be covered in this book. However, one technique that has enjoyed considerable usage in hydrological applications is *wavelet* analysis. To introduce this concept, we need to see that a period time series as defined earlier (where $x_t \approx x_{t+d}$ for *all* time) will rarely be found in reality. It is much more common that a time series is periodic over a *specific interval of time*. In which case, Fourier analysis produces some counterintuitive results. A time series which has a sinusoidal component with amplitude varying in time, for example a climate-driven variable which has year-to-year variations in amplitude (we can represent this as $A_k(t)$), will not be exactly periodic, and the associated Fourier analysis will introduce additional non-zero amplitude components, whose only role is to 'make up' the discrepancy between a precisely periodic representation, and one whose amplitude depends upon time. These additional components are not really meaningful, except in an abstract sense, and so Fourier analysis in these situations is inadequate.

Wavelet analysis is an approach to circumvent this problem. Instead of decomposing the time series in terms of completely periodic components, the components in wavelet analysis are time restricted: they have non-zero amplitude only over a specific time window. Exactly as in the DFT, we can define the so-called *discrete wavelet transform* (DWT). This allows us to do DWT analysis of a time series, which gives us a representation in terms of *both* frequency and time. It gives information not only about the amplitude of any particular periodic component, but also when, in time, that component dominates.

5.7 Smoothing and filtering

When faced with time series data, one operation might be to reduce noise in the data. This noise might occur due to the obscuring effect of errors in measurement equipment, human error, or some other random fluctuation in the observed parameter. Often, this will result in a *smoothing* operation applied to the time series. On the other hand, a *filtering* operation will isolate a particular component of the time series, for example a particular frequency component. These operations differ fundamentally to those in Section 5.3, since that the data itself is altered by the process.

Many smoothing or filtering operations have a rigorous statistical interpretation, that is they are based on a probabilistic model for the noise in the time series, and the operation leads to a statistically optimal estimate of the actual time series obscured by noise. Yet, a large number of techniques are inherently non-statistical and apply mathematical operations which have no unique statistical interpretation. Usually, the chosen mathematical operations will have some justification other than statistical.

5.8 Linear smoothing and filtering methods

Linear methods for time series smoothing and filtering are perhaps the largest group of techniques. They include the *running mean* (average), *exponential smoothing*, *low-pass filtering* and all other *frequency domain* filtering methods. The running mean is perhaps the simplest operation

$$\hat{x}_t = \frac{1}{2W+1} \sum_{i=-W}^{W} x_{t+i} \tag{5.6}$$

where $\hat{x}_t$ is the new value of the sample at time t, after applying the smoothing, for all values of t in the time series. This operation replaces the sample at position t with the average of all the samples within a window of width $2W+1$ centred around this sample. For example, $W = 1$ gives the three-sample running mean (Figure 5.5). A slight modification of this formula is if we only incorporate past history into the smoothing; the formula then becomes:

$$\hat{x}_t = \frac{1}{W} \sum_{i=1}^{W} x_{t-i} \tag{5.7}$$

with window size W. This is sometimes known as lagged (as opposed to centred) running mean.

The main property of the running mean smoothing operation is that small temporal fluctuations are attenuated, and the larger the window size, the smoother the resulting time series. Of course, there is a trade-off here – if the window size is large, the result is increasingly smoothed, but it incorporates information from a long history and future of the time series, relative to time t, which will tend to smooth away any real changes in the underlying time series. On the other hand, for small window sizes, random fluctuations will not be significantly attenuated.

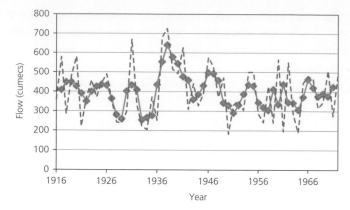

Figure 5.5 A 3-year running mean of maximum annual flows for the Morava River, shown in solid, with the original data shown as the dashed line. This is the centred running mean (mean of three values: previous, current and following) as opposed to a lagged running mean taking the previous 2 years and the current one.

A variation on the running mean filter is the *weighted running mean*

$$\hat{x}_t = \frac{1}{\sum_{j=-W}^{W} w_j} \sum_{i=-W}^{W} w_i x_{t+i} \qquad (5.8)$$

for a series of $2W + 1$ weights $w_j > 0$. These weights can be chosen to place more emphasis on samples closer to time t in the window and to place less importance on samples further into the past or future. One classical scheme which has these properties is *Gaussian weighting*:

$$w_j = \exp\left(-sj^2\right) \qquad (5.9)$$

The extent of influence of past and future samples is given by the *bandwidth parameter* $s > 0$. If this is large, the weights decay quickly with increasing temporal distance, and they decay more slowly with a small choice of bandwidth. Another linear technique which has had considerable use in time series analysis is *exponential smoothing*:

$$\hat{x}_t = ax_{t-1} + (1-a)\hat{x}_{t-1} \qquad (5.10)$$

with $\hat{x}_1 = x_1$ to start the iteration. This *recursive smoothing* forms a weighted average of the past input, and the past output of the smoothing operation, at each time step. The smoothing constant, $0 < a < 1$, determines the relative weight placed on the past input versus the past output. If a is nearly 1, then the input time series is almost completely passed through and almost no smoothing occurs. If it is close to zero, the output is based almost entirely on the past outputs, so maximum smoothing occurs.

We can get some further insight into how exponential smoothing works by rewriting it as lagged, weighted running mean. First, we set up a series of weights according to the following formula:

$$w_j = a(1-a)^j \tag{5.11}$$

for $1 < j < t$. Then, the exponential smoothing is equivalent to the following weighted running mean, with a window size that grows with t:

$$\hat{x}_t = \sum_{i=1}^{t-1} w_i x_{t+i} \tag{5.12}$$

Linear smoothing methods are special examples of *linear filtering* operations, the most common of which are *frequency-domain methods*. The approach involves computing the Fourier transform (usually by the FFT) and obtaining a set of N amplitudes α_k and phases ϕ_k for a time series of length N. The filtering operation typically involves setting some of the amplitudes to zero and then recomputing the filtered time series using the IFFT. This has the effect of removing from the original time series a selected set of frequency components. These components are said to have been 'filtered out'. Less radically, it is of course possible to attenuate or amplify, rather than completely remove, any desired frequency components.

As remarked earlier, linear smoothing is an example of linear filtering, in the following way. The effect of all linear smoothing operations can be described in the frequency domain, that is a linear smoothing operation can be completely described in terms of the amplifying or attenuating effects it has on certain frequency components. In particular, it can be shown that smoothing methods always reduce the amplitude of high frequencies, so they are known as *low-pass filters*. For example, the running mean filter is a low-pass filter whose attenuation of higher frequency components becomes more drastic as the length of the window is increased. Similarly, the exponential smoothing filter is also a low-pass filter, and the smoothing constant a determines the extent of reduction in amplitude of high-frequency components. If a is close to zero, the high-frequency components are drastically attenuated.

While the discussion of filtering is usually focused on amplitudes, the phases ϕ_k should not be entirely ignored. While we can generally ignore phase when performing frequency domain filtering, by contrast, phase is a crucial consideration in smoothing. We have seen this effect in the choice of centred versus lagged running mean: a lagged filter will introduce a time delay into the original time series, proportional to the length of the window. This time delay leads to a change in phase components. This same effect occurs with the recursive exponential smoothing filter; indeed recursive filters are in some senses much worse than non-recursive ones, because the phase changes with frequency can be difficult to control. Because of this, frequency domain filtering is usually preferred to time domain smoothing – the only advantage to time domain smoothing is that it is usually simpler to perform in practice.

There are a vast number of (considerably more complex) linear smoothing operations in use, for example the *Kalman filter* which is also often used to make predictions about future values of the time series, taking into account the autocorrelation in time and any noise in the observed data. An entirely different kind of linear filtering can be performed using wavelet analysis. Using the DWT, one can amplify or attenuate components of a certain frequency that appear simultaneously at certain points in time; the inverse DWT then allows reconstruction of the filtered time series.

In addition to smoothing and filtering of time series data, the same techniques can also be used for spatial data. One particularly important application in hydrology is for generating digital terrain models from LiDAR (light detection and ranging) data. Digital terrain models are routinely used in hydrological analysis for defining catchment boundaries, calculating slopes, generating stream networks and mapping flood extents. LiDAR is used chiefly from aircraft to capture the elevation of the ground surface. The principle is basically that time differences between the transmission and reception of a beam of light can be converted into a distance and therefore give the elevation of features on the ground. One problem though is that the beam is reflected from any object such as trees and buildings; therefore, the raw data does not give a true reflection of the ground surface. LiDAR data is presented as raster data, an extent of equal-sized square pixels (e.g. 1×1 m) each with a value pertaining to the elevation. A range of techniques are used to remove errors from the LiDAR data. Some require combination of aerial images with the LiDAR data to identify locations where the ground levels require an adjustment from the raw LiDAR data such as forested areas. Also the mathematical operations mentioned in this section are applied to remove error. A technique similar to the running mean is applied where the mean is taken for all nine cells surrounding a cell where a spurious value has been recorded. This process is graphically presented in Figure 5.6.

5.9 Nonlinear filtering methods

While linear filtering and smoothing can attenuate or amplify any set of frequency components, there are many situations in which frequency domain filtering is either inappropriate or hopelessly inefficient. An example of this is *step filtering*, that is removing noise from time series which have step changes. In this situation, the underlying time series does not change smoothly over time, and no linear filter or smoothing operation can simultaneously reduce the noise and also preserve the step changes in the underlying data. Another situation comes from statistical considerations. Consider the example of the running mean smoothing. This expects to find the best guess for the underlying time series, by computing the

(a)

(b)

Figure 5.6 The application of filtering to produce a digital ground surface model from LiDAR elevation data, where trees and other spurious elevation values in (a) (dark spots) have been removed to form the bare earth model in (b). Data from the Environment Agency geomatics group.

mean. The mean is, however, badly skewed by outliers, that is by points which are unusually far from the rest of the data. Under these circumstances, the running mean will produce bad results, and nonlinear filters are preferable.

An ubiquitous and simple nonlinear smoothing operation is the *running median*:

$$\hat{x}_t = \mathrm{median}(x_{t-W}, x_{t-W+1}, \ldots x_{t+W}) \tag{5.13}$$

This operation works by picking the middle value of the values in each running window. This is superficially similar to the running mean filter, except that it has very different properties. For example, where the running mean filter is badly skewed by outliers, up to 50% of the data in each window can be outliers and the output is still reliable. Another important property is that if the underlying signal contains a step change, the median filter will pass this through unaltered.

However, the running median cannot be naturally understood as a low-pass filter; it is a method for reducing the noise in a time series, which is much more effective than any linear filter when the errors contain outliers.

Here is a useful application for the running median. Consider the extreme values of daily rainfall, estimated as the maximum over annual blocks. These block maxima can be modelled as following a Pareto distribution (from peaks over a threshold – see Chapter 2). For some values of the parameters of this distribution, it can be shown that the mean does not exist in a formal sense, and any attempt to use running mean smoothing to get an estimate of long-term trends in extremes would be misleading. In this application, the running median is ideal because it always exists for all parameters of the Pareto.

5.10 Time series modelling

Chapter 3 introduced the concept of time within simple hydrological models where the value of a parameter at a particular time step is calculated using a value from the preceding time step. This is a common approach to modelling hydrological processes over time such as the water balance over a year. Table 5.1 shows how such a water balance model can be applied on a daily basis. Here the soil moisture deficit (SMD) function at any day (d) is the result of the soil moisture deficit function from the preceding day plus the actual evapotranspiration (AE) minus the daily rainfall function (P):

$$SMD(d) = SMD(d-1) + AE(d) - P(d) \qquad (5.14)$$

The model will assume zero soil moisture deficit for the first day, but as the crop begins to grow and take up more water, the actual evapotranspiration increases and if this exceeds rainfall then soil moisture deficit also increases. If the balance is positive then the soil moisture deficit is given as zero and any excess water is lost from the system through percolation (assuming a permeable soil).

This is an example where the model has been used to infer a consistent series of predicted measurements through time; some of the quantities would have been calculated from other models (in this case the AE quantity from the Penman–Monteith model; see Chapter 3). However, it is universally the case that if someone were to actually attempt to measure these quantities over time and put the measured quantities into the model, the results would be inconsistent. Where time series modelling has a real advantage over a pure physically based model is that it can take into account the uncertainties of observational measurements. These uncertainties mean that Equation (5.14) is only approximate. Therefore, to undertake proper time series modelling, there should be an explicit recognition that observational quantities will have a significant random component. Unlike deterministic process-based models (see Chapter 3), any realistic time series model must attempt to deal with uncertainty and, therefore, must

Table 5.1 An example of a soil water balance time series, computed using Equation (5.14), and starting with a day 0 soil moisture deficit of 0 mm.

Day	Rainfall $P(d)$ (mm)	Actual Evapotranspiration $AE(d)$ (mm)	Soil Moisture Deficit $SMD(d)$ (mm)
0			0
1	2.1	1.9	−0.2
2	5.1	1.3	−4.0
3	3.2	1.7	−5.5
4	0.6	2.1	−4.0
5	1.7	2.3	−3.4
6	1.3	2.4	−2.3
7	2.8	2.6	−2.5
8	0.6	2.9	−0.2
9	0.8	3.0	2.0
10	0.7	3.1	4.4
11	0.7	3.2	6.9
12	6.8	3.8	3.9
13	4.9	4.3	3.3
14	9.0	2.5	−3.2
15	27.0	1.6	−28.6
16	0.3	5.8	−23.1
17	2.8	6.0	−19.9
18	0.6	6.1	−14.4
19	1.3	6.2	−9.5
20	0.1	6.2	−3.4

be partly statistical. For example, a typical time series model for the equation would include a parameter a, to give an extra degree of freedom (an unknown quantity), and a random variable term $\epsilon(d)$. This is an example of a recursive linear filter:

$$SMD(d) = aSMD(d-1) + AE(d) - P(d) + \epsilon(d) \tag{5.15}$$

The parameter a would need to be estimated using some kind of regression method (see Chapter 4) and by making some assumptions on the form of the distribution of the random variable ϵ.

5.11 Hybrid time series/process-based models

We have seen in Chapter 3 the typical form of deterministic, process-based models in hydrology – they are often either algebraic, difference or differential equations. Such models represent a synthesis of various physical mechanisms such as mass conservation and empirically derived laws. Although it is entirely transparent how such models are constructed, when confronted with real hydrological time series data, they often do not make very accurate forecasts of hydrological parameters. By contrast, time series models (such as those based on prediction or smoothing using, for example, linear or nonlinear filtering) tend to outperform deterministic, process-based models in terms of predictive accuracy. The problem is that it is not at all obvious how such time series–based predictions can be related to underlying mechanistic hydrological principles.

As a very simple example of a time-series model, consider a general rainfall model which predicts rainfall depth, R, at time $t+1$ (i.e. the next time step) as a function of rainfall at time t (the current time):

$$R_{t+1} = f(R_t) + \epsilon_t \tag{5.16}$$

The function f captures all that can be inferred from the time series of rainfall depths R_t and this may well be a model based entirely on data fitting (see Chapter 4). Indeed, the soil moisture deficit time series model (5.15) is a special case of (5.16). Such time series models have been proven to be as good as rainfall forecasts derived from deterministic, processed-based numerical weather prediction models which require supercomputing resources to solve (Little et al. 2009).

Data-based mechanistic modelling (DBM) attempts to synthesise the advantages of both mechanistic, process-based modelling, using explicit representations of mechanistic principles, with the predictive forecasting accuracy of time series modelling. DBM hydrological forecasting models include the *state-dependent parameter rainfall-runoff-flow routing catchment method* (Young 2002). This method includes a nonlinear component that models the mechanistic principles involved in catchment storage effects to calculate the effective rainfall (U):

$$U_t = f(R_t, Q_t, E_t, T_t) \tag{5.17}$$

where the effective rainfall for the current time step t is a factor of rainfall, flow, evaporation (E) and temperature (T).

The flow is predicted from a linear combination of past sampled values of itself and past and present sampled values of the effective rainfall, which is a recursive linear filter that represents the catchment response in terms of the stream flow (Q) to an effective rainfall impulse:

$$\begin{aligned} Q_t = -a_1 Q_{t-1} - a_2 Q_{t-2} - \cdots - a_N Q_{t-N} \\ + b_0 U_{t-d} + b_1 U_{t-d-1} + \cdots + b_M U_{t-d-M} + \epsilon_t \end{aligned} \tag{5.18}$$

Here, the parameters d, $a_1, a_2 \ldots a_N$ and $b_0, b_1 \ldots b_M$ are constants which are to be determined by some kind of model fitting procedure and ϵ_t is a noise term to account for measurement error, the effects of unmeasured inputs and modelling error.

Data-based mechanistic models (5.17 and 5.18) have had some practical applications particularly in the role of real-time flood forecasting, that is forecasting flood flows or levels in a river when the flood event is actually happening.

The disadvantage though is that they require large amounts of observed data and a high level of mathematical sophistication in order to find 'good' nonlinear functions f and to develop appropriate regression methods to estimate the parameters in (5.18). That also can lead to the problem that such models are not often challenged by the hydrological community due to the complexity and abstract nature of the mathematics.

There is a further problem that time series approaches are heavily reliant on past data records, so that any changes to the method of observations, monitoring equipment or at a more general level within the catchment (e.g. land use change) would have significant impacts on the model predictions. For example, if a DBM model was developed based on water level observations over the past 50 years, but recent engineering works such as a weir enlargement near the monitoring station had caused an overall drop in water levels, the model predictions would be completely invalid. The problems of how data records may change over time is dealt with in the following section on non-stationarity.

5.12 Detecting non-stationarity

A stationary process, by definition, is a process where the probability distribution of the measured variables at any set of points in time is independent of time (WMO/UNESCO 1992). The problem of *non-stationarity* arises in hydrology where a factor or factors which influence the process have sudden or gradual changes over time. Sudden changes, as described in the previous section, can be the modification of a hydraulic structure which would affect flows and water levels on a river or drastic changes in land use in a catchment such as the development of a rural area into an urban environment with impermeable surfaces.

Gradual changes may result from a less drastic change in land use or land management such as the application of new agricultural practices or vegetation change, but by far the greatest concern in hydrological work over the past few decades has been the influence of climate change.

Sudden changes, particularly relating to modifications of the gauging station or river channel, should be properly documented and understood by the monitoring authorities. This will allow the necessary adjustment factors to be implemented so that the observations pre- and post-change can be directly compared. Some gradual changes can also be implemented reasonably accurately, for example when looking at water levels in tidal reaches, the known oscillations of the moons orbit (the 18.6 year lunar cycle) can be factored into historical observations and in some locations the effect of falling or rising sea levels can also be incorporated. This last point is particularly relevant to Scandinavia where the crust is still rising from isostatic readjustment following the melting of the ice sheets (Figure 5.7).

Other gradual changes are more difficult to assess, particularly in relation to the natural variability of the observed hydrological data such as river flows.

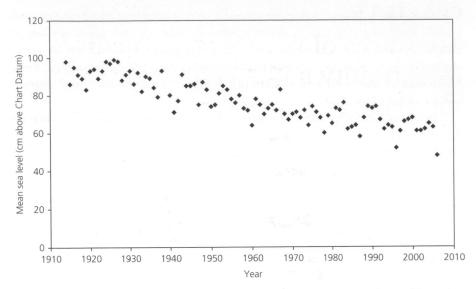

Figure 5.7 Mean sea levels for Oslo, 1910–2005. Data from the Norwegian Hydrographic Service.

In such cases, time series tests are used to establish whether a sampled process is stationary or non-stationary. For example, if the non-stationarity is expected to appear in the form of trends, one can use the statistical techniques discussed in Section 5.4.

References

Cover, M. & Thomas, J.A. (2006) *Elements of Information Theory*. Wiley Interscience, New York.

Little, M.A., McSharry, P.E. & Taylor, J.W. (2009) Generalized linear models for site-specific density forecasting of UK. *Daily Rainfall, Monthly Weather Review*, **137** (3), 1031–1047.

Rodda, H.J.E. & Hawkins, J. (2012) Surface water flooding in rural areas: observations, monitoring and methods for peak flow estimation. Proceedings of the British Hydrological Annual Symposium, Dundee, July 2012.

UNESCO (1976) *World Catalogue of Very Large Floods*. UNESCO Press, Paris.

Wasserman, L. (2003) *All of Statistics: A Concise Course in Statistical Inference*. Springer, New York.

WMO/UNESCO (1992) *International Glossary of Hydrology*. World Meteorological Organization, Geneva.

Young, P.C. (2002) Advances in real-time flood forecasting. *Philosophical Transactions of the Royal Society of London. Series A*, **360**, 1433–1450.

CHAPTER 6

Measures of model performance, uncertainty and stochastic modelling

6.1 Introduction

Today, most hydrological studies involve modelling to predict some quantity, timing or magnitude of a given parameter. The result of such modelling needs to be expressed in a way that demonstrates how well the model has performed. A so-called 'goodness of fit' is often calculated as a measure to assess model performance, such as in the Flood Estimation Handbook software (Institute of Hydrology 1999). The need to assess model performance is more evident when studies have used a number of different models for the same problem and the outputs are then used to guide policy decisions. For example, climate change studies often consider scenarios using up to 10 different global circulation models. The performance of a model is not usually confined to predicting a single output but more often outputs at a range of study sites, or outputs for a single site for a number of events, or a combination of spatial and temporal distributions. Therefore, the question is which set of predictions are better of the two model predictions shown in Figure 6.1.

6.2 Quantitative measures of performance

Common measures of model performance include the *mean square error* (MSE), the *root mean square error* (RMSE), *mean absolute error* (MAE), *maximum error* (ME), the *correlation coefficient* (R^2), and the *coefficient of residual mass* (CRM). Many of these measures are based around the idea of adding up some function of the difference

Understanding Mathematical and Statistical Techniques in Hydrology: An Examples-Based Approach, First Edition. Harvey J. E. Rodda and Max A. Little.
© 2015 Harvey J. E. Rodda and Max A. Little. Published 2015 by John Wiley & Sons, Ltd.

between the observed data x_i and the predicted data y_i, over the N available data points, here indexed as $i = 1, 2 \ldots N$. Below, we list a few example measures:

$$\text{MSE} = \frac{1}{N} \sum_{i=1}^{N} (x_i - y_i)^2 \tag{6.1}$$

$$\text{RMSE} = \sqrt{\frac{1}{N} \sum_{i=1}^{N} (x_i - y_i)^2} \tag{6.2}$$

$$\text{ME} = \max_{i=1,2 \ldots N} |x_i - y_i| \tag{6.3}$$

$$\text{MAE} = \frac{1}{N} \sum_{i=1}^{N} |x_i - y_i| \tag{6.4}$$

$$R^2 = \frac{\left[\sum_{i=1}^{N} (x_i - \bar{x})(y_i - \bar{y}) \right]^2}{\sum_{j=1}^{N} (x_j - \bar{x})^2 \sum_{k=1}^{N} (y_k - \bar{y})^2} \tag{6.5}$$

$$\text{CRM} = \frac{\bar{x} - \bar{y}}{\bar{x}} \tag{6.6}$$

where the overbar ($\bar{x}$) indicates the mean. The equations can be described in words as follows:

MSE. The difference between each of the predicted and observed values squared to remove the effect of positive and negative values, then summed (giving the sum of the squared difference) and divided by the number of values to get the average.

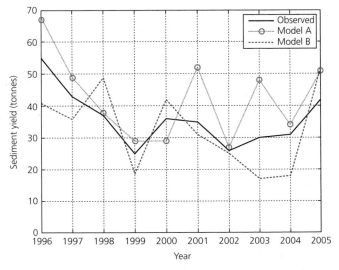

Figure 6.1 Hypothetical model predictions from two different sediment load models compared with observed sediment load for a catchment.

RMSE. The difference between each of the predicted and observed values squared to remove the effect of positive and negative values, then summed (giving the sum of the squared difference) and divided by the number of values to get the average and finally given as the square root to remove the effect of squaring the values.

ME. The maximum absolute difference (as indicated by the straight vertical bars rather than brackets) between each of the observed and predicted values.

MAE. The absolute difference between each of the predicted and observed values summed, giving the sum of the difference, and divided by the number of values to get the average.

Squared correlation coefficient (R^2). The square of the covariance divided by the product of the variance for x and y where the variance (as introduced in Chapter 1) is the average squared difference between each sample and the mean of the sample set. The covariance is the average product of the difference between the two variables and their corresponding means. This measure is commonly referred to as the *R*-squared value.

CRM. The difference between the mean of all the observed values and the mean of all the predicted values divided by the mean of all the observed values.

It would be understandable for non-mathematicians to get quickly confused by the nature of the computations introduced previously. Important questions need to be answered before selecting one or more of these measures to make model performance comparisons. These include questions such as the following: what value will the measure take if the predictions are perfect? What is the numerical range of the measure? How does the numerical range of the measure compare to the range of the data? How does the measure respond to one very bad prediction amongst otherwise good ones? These questions are addressed in the following texts.

6.3 Comparing measures

MSE and RMSE. Both measures are based on the concept of taking the square of the difference between the actual and predicted data. Therefore, MSE and RMSE cannot be negative. In both cases, the average square difference over all data points is calculated. Since the MSE takes the square of the data points, it does not share the same numerical scaling as the data. By contrast, since the RMSE 'undoes' the squaring operation, it has the same numerical scaling as the observed data. For this reason, model errors assessed using RMSE can be interpreted with respect to the range of the data. If the predictions are perfect, then both MSE and RMSE are zero. Since both MSE and RMSE use the square difference, one very large error contributes very significantly to the sum. This means that MSE and RMSE are not *robust* to the occasional bad prediction.

ME and MAE. As with MSE and RMSE, these measures calculate the absolute difference between the observed and predicted data. As such neither measure can be negative. ME returns the maximum absolute value of the difference over all the data points, whereas MAE returns the mean absolute difference. If the predictions are perfect, then both ME and MAE are zero. ME and MAE differ dramatically in response to single, large prediction errors: MAE is not adversely affected by them, whereas ME is entirely determined by the largest prediction error. Of MAE, RMSE and ME, MAE is the least affected, where ME is the most affected. Both ME and MAE are on the same scale as the data.

R^2 is a *correlation measure*. It ranges between 0 and 1. When 1, it indicates that there is a perfect linear relationship between the predictions and the data. That is, if we plotted the observations on the *x*-axis, and the predictions on the *y*-axis, all the points would lie on a perfect straight line. If the measure is zero, it indicates that there is no linear relationship between the predictions and the observations. Unlike the previous measures, this measure does not tell us whether the predictions are *unbiased* or not – the predictions could be very much larger or smaller than the observations but still could be highly correlated.

CRM. This measure quantifies the difference in means of the predictions versus the observed data, relative to the mean of the observations. If the mean of the predictions is the same as that of the observed data, then CRM is 0. If the mean of the predictions is 0 on the other hand, then CRM is 1. Note that CRM is undefined if the mean of the observations is 0.

A list of the results of each of these measures for models A and B is presented in Table 6.1. What should be noted is which model comes out best depends on the choice of performance measure. Therefore, unless there are compelling reasons to use one particular performance measure over another, it is always worth computing several to see whether there is any pattern of agreement or not. In particular, in this case, model B seems to win out more often than model A when using different performance measures, so it could be argued overall, that model B is more accurate than model A.

Some interesting observations can be drawn from Tables 6.1 and 6.2. In particular, we can see that, according to the RMSE and MAE, the average prediction error is around 25% of the average of the observations, whereas the maximum error is around 40% of the average of the observations (we would expect the maximum error to be the largest of similar error measures). Also as expected,

Table 6.1 Comparing results of measures quantifying the accuracy of predictions from models A and B, for the catchment sediment load data shown in Figure 6.1.

Performance Measure	Model A	Model B	Winner	Winning Criteria
MSE	95.00	91.60	Model B	Closer to 0 is better
RMSE	9.75	9.57	Model B	Closer to 0 is better
ME	18.00	14.00	Model B	Closer to 0 is better
MAE	6.80	8.60	Model A	Closer to 0 is better
R^2	0.66	0.45	Model A	Closer to 1 is better
CRM	−0.18	0.08	Model B	Closer to 0 is better

Table 6.2 Statistical properties of the catchment sediment load data and predictions shown in Figure 6.1.

Statistical Quantity	Observed Data	Model A Predictions	Model B Predictions
Mean	36.0	42.4	33.0
Maximum	55.0	67.0	52.0
Minimum	25.0	27.0	17.0
Variance	81.1	170.3	168.4
Standard deviation	9.0	13.0	13.0
Median	35.5	43.0	33.5

See Chapter 1 for definitions of these statistical quantities.

the MAE and RMSE are quite close. It is also interesting to note that the correlation coefficient is significantly higher for model A than model B, although there is not really enough data to distinguish this difference from a chance occurrence.

6.4 The Nash–Sutcliffe method

Another measure of model performance was specifically developed by hydrologists for assessing estimates of flows. This is the *Nash–Sutcliffe method* which calculates a model *efficiency E* based on the difference between modelled and observed flows and modelled and mean flows as shown below:

$$E = 1 - \frac{\sum_{i=1}^{N} (Q_i - \hat{Q}_i)^2}{\sum_{j=1}^{N} (Q_j - \bar{Q})^2} \tag{6.7}$$

where Q_i is the observed flow at time i, $\hat{Q}_i$ is the modelled flow at time i and Q is the mean of the observed values. The right-hand side of this equation is given the term F, which can be understood as the ratio of the MSE to the variance of the observed data – how large is the MSE by comparison to the variance of the observed data. If the MSE was equal to the variance, then F would be 1 but the MSE could be much larger than the variance and also the MSE cannot be smaller than zero. Therefore, the quantity F lies between zero and plus infinity. The quantity E is $1 - F$, so if F was 0 E would be 1, alternatively if F was infinity E would be negative infinity. If F is equal to 1 then E is 0. In fact, the formula (6.7) is just a linear rescaling of the MSE, but it is widely used in hydrology.

Values closer to 1 indicate better performance, a value of 0 indicates that the model predictions are only as good as the mean of the observed data, and values below zero indicate that the model is poor with the mean of the data being closer to the observed values than the model predictions. Computations required to calculate E are demonstrated for hypothetical data in Table 6.3.

Table 6.3 Example computation of the Nash–Sutcliffe efficiency for a set of observed and modelled data, using Equation (6.7).

Sample	Observed Q_i	Modelled $\hat{Q}_i$	$Q_i - \hat{Q}_i$	$(Q_i - \hat{Q}_i)^2$	$Q_j - \bar{Q}$	$(Q_i - \bar{Q})^2$
1	55	37	18	324	19	361
2	43	27	16	256	7	49
3	37	35	2	4	1	1
4	25	30	−5	25	−11	121
5	36	29	7	49	0	0
6	35	29	6	36	−1	1
7	26	25	1	1	−10	100
8	30	23	7	49	−6	36
9	31	34	−3	9	−5	25
10	42	25	17	289	6	36
Mean	$\bar{Q} = 36$					
Sums				1042		730
Efficiency E	−0.43					

Unfortunately for this example, the model efficiency is poor (<0), and from looking at the data, it appears that most of the values are underestimates.

6.5 Stochastic modelling

Stochastic modelling approaches typically apply the theory of probability and random sampling to deterministic process models by not simply having a single value for each model parameter but having a range of values.

The International Glossary of Hydrology (WMO/UNESCO 1992) defines stochastic hydrology as 'hydrological processes and phenomena which are described and analysed by the methods of probability theory'. The word stochastic is actually derived from a Greek word which means to aim or shoot an arrow at a target (Koutsoyiannis 2000). In the field of modelling, the term stochastic is synonymous with 'random' or 'probabilistic', and for hydrology in particular, it is one of a variety of descriptions used for different characteristics which the model may possess. For example, models can be referred to as *lumped* or *distributed* if they consider the catchment as a single homogeneous unit or if different parts of the catchment are treated differently due to their hydrological characteristics. Also, models can be described as *physically based* if they have equations which simulate the physical processes which are modelled (such as those described in Chapter 3) or *empirical* if the equations driving the model are based on the fitting of relationships to data (Chapter 4). A third tier of classification groups models as either *deterministic* if the output is a fixed single value or *stochastic* if a range of values with a given probability are generated.

6.6 Monte Carlo simulations

In order for a deterministic process model to generate a range of outputs with a given probability, there also needs to be a range associated with one or more of the model input parameters. The easiest way to provide this range is to take an upper and lower limit of the parameter, often from the knowledge of observed values, and then to generate a random number uniformly within this range for each simulation. Uniform random number generation is often a basic component of computer programs or packages, and when it is repeated a large number of times (many thousands, typically), the model outputs will fall into an observed distribution which can then be estimated. This technique is often known as the *Monte Carlo* method; the origin of the name is the codename given to the technique first used in modelling radiation shielding by scientists at the Los Alamos Scientific Laboratory. The name reflects the use of chance but also the casino in Monaco where one of the scientists' relatives was known to gamble.

The Monte Carlo method is often applied to situations where a combination of parameters is known to cause a given output. For example, dam design is governed by the need to find the most severe flood which could take place in a catchment, and the most severe flood would be a combination of extreme rainfall and the antecedent conditions. For an area where flooding is caused by spring snowmelt, the two parameters of most concern would be the rainfall for a given duration (e.g. 2 days) over the spring period and the depth of snow which is ready to thaw. Historical observations from the selected catchment and neighbouring within a wider region can be used to define the range of 2 day rainfalls, for example 0–250 mm, and maximum snow depths, for example 0–200 cm.

Using uniformly generated random numbers from these ranges, combinations of parameter values can be generated (see Table 6.4, where for a sample of

Table 6.4 Example uniform random number combinations within known ranges of rainfall and snow depth.

Event	Rainfall (mm)	Snow Depth (cm)
1	99	58
2	188	16
3	100	59
4	152	90
5	77	139
6	7	127
7	26	158
8	24	101
9	211	145
10	74	169

The most extreme joint combination of these two parameters is highlighted.

10 random number combinations the worst case is highlighted) and used as inputs into a hydrological model. For many thousands of simulations, it might be expected that the worst case combination of maximum rainfall and maximum snow melt would eventually be sampled. However, it is important to appreciate that because of the complex nature of the hydrological model there may not be a simple increasing or decreasing relationship between rainfall, snow depth and the model output. When more parameters are included with a uniform random selection from a range of values, where the effect on the overall output is not known, then the application of the Monte Carlo method becomes more worthwhile and can provide good understanding of how a particular system will respond to changing parameter values.

The outputs from a stochastic model, can then be plotted as a histogram and a probability distribution estimated by normalizing the histogram (i.e. dividing the histogram counts through by the number of simulations). This would then give a probability associated with each model prediction, based on the range of the input parameter values. This provides a more nuanced alternative to a simple deterministic output which might represent only the worst case (e.g. maximum rainfall and snow depth) but without any idea of how likely this worst case event may be.

The stochastic modelling may also suggest a plausible range of outputs which could be used to inform decisions based on the range of potential output magnitudes. For example, a dam design might need to be able to withstand a 1 in 10,000 year flood, to ensure a downstream population was at a minimal risk of any catastrophic flooding, but for a less vulnerable component such as an access road, it would be sufficient for it to be safe up to the 1 in 100 year flood level. By providing designs which cater to the worst case scenario, an unnecessary excess cost would be required which might not be a desirable trade-off for the degree of risk.

It should be pointed out that this basic Monte Carlo method has some serious limitations. For example, it does not take into account any genuine dependence which might exist between parameters. In the example discussed previously, snow depth and rainfall may well be related so that a simple simulation that assesses the probability of the model output and does not respect the *joint* probability between these two parameters could be unrealistic. Another limitation is that as the number of parameters increases, the number of simulations must grow *exponentially* with this increase in parameters, in order to sample the range of possible output parameters with uniform precision. This is necessary in order to provide a reliable estimate of the probability distribution of the model output. For example, if, with two parameters, the sufficient number of simulations is 1,000, then with three parameters, the number of simulations should be on the order of 10,000 and with four parameters it should be around 100,000. This is not a problem if the model is simple, but for more complex models, the computational requirements can become prohibitive.

6.7 Non-uniform Monte Carlo sampling

A refinement to uniform Monte Carlo random sampling is used particularly in studies which aim to consider the most extreme events, for example in the field of *catastrophe modelling*. In the past decade or so, catastrophe models have been developed for the insurance sector which aim to predict the potential damage and financial loss associated with natural hazards such as hurricanes (and other windstorms), earthquakes and floods. Such models are based on a regional or countrywide scale so that the output is, for example, the 1 in 100 year financial loss from flooding for the whole of the Czech Republic or the 1 in 100 year financial loss in California expected from earthquakes.

The objective of such models is to provide a combination of the most extreme values for the key driving parameters so that the worst case scenario is simulated. The ranges of parameters are decided using observations from historical events or knowledge of the physical processes involved. The analysis of historical events as part of the development of such models can be very useful. For example, the 1953 storm surge which caused considerable damage and loss of life in the United Kingdom, the Netherlands and Germany was the result of an intense depression coinciding with high spring tides. In some locations, however, the storm surge hit the shore when the tide was receding, so for the purposes of a catastrophe model, events would need to be generated where the increased water level due to the storm surge would coincide with the maximum tidal level.

Within the catastrophe model framework, in order to ensure that an event is generated where the maximum parameter ranges are encountered, a large number of synthetic events must be produced. This requirement, coupled with the regional or countrywide scale of the models means that relatively less attention is given to the details of the scientific processes when compared to models designed for a single location. In order to sample events which have the most severe combination of parameters, a *non-uniform* sampling method is used. Many approaches could be used, including sampling from extreme value distributions (Chapter 2), but we will focus on the simple approach known as *stratified sampling* which has the virtue of being very easy to describe (we should point out that this name is used to describe other statistical techniques in different contexts and so should not be confused with the use outside the context of catastrophe modelling).

Instead of using uniform random numbers to choose parameters values within the full possible range (e.g. a 2-day rainfall between 0 and 250 mm), a greater number of samples are taken at the higher end (e.g. 200–250 mm) to ensure that a larger number of the more extreme parameter combinations are sampled. The same number of samples can be generated but instead of using, for example 1000 values chosen uniformly between 0 and 250, the sampling is broken down into a number of intervals or slices (e.g. 10), where the same number of samples are taken from within each slice (100). The slices that can be set up to ensure more

Table 6.5 Example slices for 2-day rainfall ranging from 0 to 250 mm, using non-uniform random stratified sampling to emphasize extremes.

Sample Slice	Range	Percent of Range
1	0–125	50
2	126–150	10
3	151–175	10
4	176–195	8
5	196–210	6
6	211–220	4
7	221–230	4
8	231–240	4
9	241–245	2
10	246–250	2

values are taken from the higher end, as shown in Table 6.5. In this way, there should be many more values within the top 10% of the distribution (i.e. maximum of the range of values). Effectively, we define, and sample from, a non-uniform distribution over the range of values which places higher probability on the more extreme values of the range.

Stratified sampling is often used to develop stochastic models where the impact of a combination of different parameters on determining the most extreme values is not always clear. It would benefit the spring flood example shown earlier, if further factors which affected the rate of snow melt such as air temperature and wind speed were included as model parameters. The combination of the parameter values which would bring about the greatest flood should therefore be identified when enough samples are taken for the uppermost 10% of the values, as an example. The frequency of the extremes which are generated from this type of modelling is much greater than from uniform sampling as shown in Figure 6.2. To compensate, however, catastrophe modellers would assume that the events generated from the model would be over a much larger time period (e.g. 10,000 years) than the time period used for defining the range of input parameter values which may be less than 100 years.

Some care must be taken so that parameters are sampled in ways which will ensure the most severe outputs are based on a physical understanding of the system. For example, in the case of a river flood catastrophe model from extreme rainfall, a more severe scenario would result from the combination of a longer duration of the rainfall event and a slower translational speed of the weather system which brings about the rainfall. The sampling of the translational speed should effectively be reversed so that more samples are taken in the lower extremes of its values. Some parameters may also be dependent on other parameters and hence not included within the sampling. In this case, the dependent parameter is simply calculated based on the value of the controlling parameter.

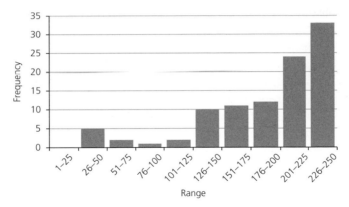

Figure 6.2 A histogram derived from stochastic stratified sampling using the data ranges and sampling slices (10 values per slice) as shown in Table 6.5.

This form of non-uniform stochastic modelling can rapidly generate a set of many thousands of extreme events which would normally be experienced over a very long time period, for example 10,000 years. The advantage for considering extreme events is that through this method it is easy to define the return period – the greatest magnitude flood over the 10,000 year period is simply the 1 in 10,000 year flood, and there is no need to apply the more complicated statistical methods for extracting extreme values from much shorter records as discussed in Chapter 2.

Non-uniform sampling helps to alleviate the problem of the exponential rise in demand for simulations when combined parameters need to be sampled by focusing sampling precision in the ranges of most interest. However, it does not fully eliminate the need to assess how well the eventual model output range has been sampled such that it is possible to estimate reliable probabilities for ranges of the model output variables.

One of the characteristic of stochastic modelling is the use of random numbers being used to select input parameter values over a range of possible outcomes. The approach is also applied in some of the methods used to deal with the uncertainty associated with modelling. Here the aim is not to generate a large number of potential outputs but instead to consider the sensitivity of the various model parameters and optimize the parameter values, as described in the following section.

6.8 Uncertainty in hydrological modelling

The appreciation of uncertainty in deterministic process modelling (see Chapter 3) is a fairly recent development in hydrology, only coming to the fore in mainstream application over the past decade or so but its importance had been identified in research work in the 1990s. In early hydrological models, the developers and users were probably just glad to have a working model which gave reasonable outputs in a readily available format which could be understood by managers and policy makers, so uncertainty was often overlooked.

Uncertainty is now widely reported in the scientific literature relating to hydrological modelling and other modelling of natural processes. It can be thought of as the next step of a study after running a model and using statistical measures of the model performance. Where the model has not performed particularly well, the concept of model uncertainty can be used to identify ways to improve the model. Beyond the fact that uncertainty is probably most easily expressed using the mathematical framework of probability (see Chapter 1), there is in general no agreement in the hydrological community about what it means in terms of a single, concrete method which can be used to make calculations.

The uncertainty associated with the use of models to predict and estimate the magnitude and timing of a future event can arise from a variety of sources. The most widely identified uncertainties are (i) measurement uncertainties, (ii) data uncertainties, (iii) modelling uncertainties and (iv) natural variability uncertainties. Out of these, measurement and data uncertainties refer to the techniques used to measure the parameters which may have error associated with them, and how the data is handled and formatted. Data uncertainty is typically higher if the record length is short or incomplete and also whether any additional factors may change the record over time, such as any non-stationarities (Chapter 5). Uncertainty related to modelling can be attributed to the choice of statistical techniques, physical processes and simulation methods. In *all* models, some assumptions are made to simplify reality and hence this causes uncertainty. Finally the uncertainty associated with the natural variability of a system such as the hydrological cycle is another source, which can be quite often observed in work which considers trends over time, such as climate change (Chapter 5).

The use of specific methods to provide an assessment of the uncertainty associated with a modelling study is now quite common within the hydrological community and such methods are readily encountered in the scientific literature.

6.9 Uncertainty in combined models

The stochastic approach to modelling is explicitly designed to address the problem of model parameter uncertainty. In a complex model such as a probabilistic flood catastrophe model, the estimated magnitude of the flood is likely to be the output of a number of combined models each with their own associated sources of uncertainty:

1 Rainfall model – used to generate a number of rainfall scenarios
2 Rainfall–runoff model – used to derive river flows resulting from the rainfall
3 Flood routing model – used to calculate flows in downstream rivers outside the area affected by the rainfall

4 Hydrodynamic model – used to calculate flood water levels in the river channel based on the calculated flows and the morphology of the channel

5 Flood extent model – used to propagate the flood water across the floodplain and define flood depths based on the river flood levels and characteristics of the land surface

At each step, there is the uncertainty associated with the calculations and that uncertainty is passed onto the next step. If the model was providing a purely deterministic output then the uncertainty associated with the ultimate model prediction could be very high. However, by running large numbers (e.g. many thousands) of scenarios and giving outputs on a relatively coarse level (e.g. flood depth to the nearest 0.1 m per postcode), a distribution of results is provided as the output so that users can themselves gauge the overall uncertainty.

6.10 Assessing uncertainty given observed data: Bayesian methods

We have seen above that stochastic modelling, in the absence of real data, allows an assessment of uncertainty by random sampling from chosen parameter ranges. So, there is an implicit assumption that both the model parameters and the model output are treated as random variables, and this is a central assumption of the Bayesian approach (see Chapter 2). It is therefore a simple question to ask: if we do actually have some real data to calibrate the model output, how could we estimate the *posterior* distribution of the parameters given this real data?

The obvious statistical solution to this problem involves writing down a full set of probability distribution functions for the parameters, the data conditioned on the parameters and the parameters given the data. In the Bayesian terminology: the prior, the likelihood, and the posterior respectively. This is indeed possible in simple cases, but the mathematical expressions become impractical for any hydrological process model of sufficient complexity. In more realistic hydrological applications, it is not really the prior distributions which are problematic, for example we can choose this to be very simple (and often do – typical stochastic modelling approaches assume 'the prior' is uniform). It is instead the likelihood that is problematic and also computing the distribution over the data required to apply Bayes rule (it is the denominator in Eq. (2.7)).

A special technique, *generalized likelihood uncertainty estimation* (GLUE), developed by hydrologists (Beven and Binley 1992) has been proposed and is widely used to provide distributions of parameter values when data to calibrate the model output is available. The approach avoids the need to write down a complex likelihood density function for the model and instead using random sampling of the parameters to generate sets of model outputs which are then compared to the real data. From this comparison, a set of plausible model parameter values are obtained, and these plausible parameter values are used to estimate their distribution given the real data, that is an estimate of the posterior. The comparison is based on an arbitrarily chosen function which combines the parameters with the data, called a *generalized likelihood* function $L(a, x)$, where a is a vector of parameters and x is a data vector.

Here are the steps in the GLUE procedure as presented by Nott et al. (2012):

1 Pick prior distributions for the parameters (often uniform, i.e., choose ranges for the parameters) and generate a series of samples $a_1, a_2 \ldots a_N$ for these parameters.
2 For each sampled parameter value a_i, compute $L(a_i, x)$.
3 Choose an 'acceptability threshold' c and throw away all parameters a for which $L(a_i, x) < c$, retain the rest, and we will denote the remaining sampled parameters as a_k for $k = 1, 2 \ldots K$.
4 Compute the weights $w_k = L(a_k, x) / \sum_{k'=1}^{K} L(a_{k'})$.
5 The distribution of any random variable of interest (e.g. the parameter a or the model output) is the probability vector w_k associated with the parameter value sampled on run k.

The generalized likelihood function L is key to the method. Typically, it is chosen to be non-negative and to be small for model outputs which differ significantly from the measured data. In particular, RMSE^{-T} where $T > 0$ has this property (Nott et al. 2012). Because GLUE uses random sampling to simulate possible model outputs, it has a lot in common with stochastic sampling methods described earlier, but it is essentially a parameter estimation method. In fact, it has been demonstrated that GLUE is a special kind of *approximate Bayesian computation* (ABC) method (Nott et al. 2012). ABC methods are widely used to produce Bayesian posterior parameter distribution estimates where the likelihood function is complex, as often happens in physical modelling contexts.

An example of practical usage can be found in Wang et al. (2006) who applied the method to fit subsurface flow parameters using the DRAINMOD model. In this case, the following generalized likelihood was used:

$$L(a_i, x) = \exp\left(-\frac{\text{MSE}_i}{\min(\text{MSE})} \right) \tag{6.8}$$

which can be shown to be similar to the classical likelihood function arising assuming Gaussian errors in the data.

The appeal of the GLUE method is the simplicity and freedom to choose the generalized likelihood in such a way that does not restrict the choice of physical process models. There are, however, many pitfalls for the unwary; for example, the likelihood function (6.8) is not *concave* (that is having a single bump), so that a unique posterior mode does not actually exist. Stedinger et al. (2008) point out some of the deficiencies of the method, most crucially that choosing arbitrary likelihoods which do not match the real distribution of the error in the data can lead to results that are statistically invalid. Caution is therefore required in departing from the classical framework of probabilistic modelling and Bayesian reasoning.

References

Beven, K.J. & Binley, A.M. (1992) The future of distributed models: Model calibration and uncertainty prediction. *Hydrological Processes*, **6**, 279–298.

Institute of Hydrology (1999) *The Flood Estimation Handbook*, **5** Volumes. NERC Publication, Wallingford.

Koutsoyiannis, D. (2000) A generalized mathematical framework for stochastic simulation and forecast of hydrologic time series. *Water Resources Research*, **36** (6), 1519–1533.

Nash, J.E. & Sutcliffe, J.V. (1970) River flow forecasting through conceptual models part I – a discussion of principles. *Journal of Hydrology*, **10** (3), 282–290.

Nott, D.J., Marshall, L. & Brown, J. (2012) Generalized likelihood uncertainty estimation (GLUE) and approximate Baysian computation: What's the connection? *Water Resources Research*, **48**, W12602.

Stedinger, J.R., Vogel, R.M., Lee, U.S. & Batchelder, R. (2008) Appraisal of the generalized likelihood uncertainty estimation (GLUE) method. *Water Resources Research*, **44**, W00B06.

Wang, X., Frankenberger, J.R. & Kladviko, E.J. (2006) Uncertainties in DRAINMOD predictions of subsurface drain flow for an Indiana silt loam using the GLUE methodology. *Hydrological Processes*, **20**, 3069–3084.

WMO/UNESCO (1992) *International Glossary of Hydrology*. World Meteorological Organization, Geneva.

Glossary

Correlation: A measure of the extent to which a change in one variable tends to correspond to a change in the other. Linear dependence is given by the correlation coefficient ρ. If variables are uncorrelated random variables, then $\rho = 0$. Values of +1 and −1 correspond to full positive and negative correlation, respectively. Note: The existence of some correlation need not imply that the link is one of cause and effect.

Cumec: Abbreviation for cubic metres per second, the standard unit for the flow of water.

Cumulative distribution function (CDF): The probability of a random variable taking a value up to and including a certain threshold value.

Dependence: The extent to which the distribution of a random variable depends upon another.

Extreme value theory: The branch of statistics dealing with the extreme (rare) values of a probability distribution. It seeks to assess, from sample data from a given random variable, the probability of events that are more extreme than any observed value.

Frequency: The expected number of times that a particular event will be observed within a specific time frame.

Gamma distribution: A two-parameter distribution on the positive real line.

Gaussian: A two-parameter distribution defined on all real values. Usually parameterized by mean and standard deviation (or variance), it has one mode and is symmetric about that mode.

Joint probability: The probability of specific values of two or more variables occurring simultaneously.

Mean: The expected or average value of a random variable. This is also the first moment.

Median: The value of a random variable for which 50% of the distribution is above and below this value.

Mode: The largest probability value of a random variable.

Moment: The expectation of a random variable raised to the power of k. The special case $k = 1$ corresponds to the mean.

Normal distribution: Another name for the Gaussian.

Understanding Mathematical and Statistical Techniques in Hydrology: An Examples-Based Approach, First Edition. Harvey J. E. Rodda and Max A. Little.
© 2015 Harvey J. E. Rodda and Max A. Little. Published 2015 by John Wiley & Sons, Ltd.

Probability (and probability of exceedance): A dimensionless number that is always at least 0, and at most 1. A probability of 0 indicates that the event is impossible and cannot occur, and probability 1 represents an inevitable occurrence. Probability can be expressed as a fraction, percentage or a decimal. For example, the probability of obtaining a six with a shake of a fair dice is 1/6, 16.6% or 0.166. In the context of extreme flows the annual probability of exceedance is used to express the chance that a particular event will be equaled or exceeded within a given year.

Probability density function (distribution) (PDF): A function that describes the distribution of a random variable. Areas under this distribution are probability values.

Random variable: A function that maps random events to integers or real values, labeling these events with a specific number. Random variables can be either discrete (can take one of a countable, but potentially infinite, set of values) or continuous (can take any real value).

Rank: When a list of numbers is sorted, the rank is the position of any number in that sorted list.

Standard deviation: The square root of the second central moment of a random variable, that is, the square root of the second, $k = 2$, moment of the variable with the mean removed first.

Stationarity and non-stationarity: 'Stationarity' refers to the constancy of the laws and processes that govern a response of interest (e.g. flow). A stationary stochastic process has the property that its probability distribution does not change with time. That is, if parameters such as mean and variance exist, they are constant. In contrast, statistical properties of a non-stationary process vary over time. For example, abrupt or gradual changes in the mean, variance, higher moments, or characteristics of extremes may be observed.

Uncertainty: A general concept that reflects a lack of sureness about something, ranging from just short of complete sureness to an almost complete lack of conviction about an outcome. Two forms of uncertainty are often classified (i) aleatory from natural variability and (ii) epistemic from a lack of knowledge.

Variable: A mathematical quantity that can change.

Variance: The second central moment of a random variable, that is, the second moment of the variable with the mean removed first.

Index

Understanding Mathematical and Statistical Techniques in Hydrology: An Examples-Based Approach, First Edition.
Harvey J. E. Rodda and Max A. Little.
© 2015 Harvey J. E. Rodda and Max A. Little. Published 2015 by John Wiley & Sons, Ltd.